FREE Test Taking Tips DVD Offer

To help us better serve you, we have developed a Test Taking Tips DVD that we would like to give you for FREE. **This DVD covers world-class test taking tips that you can use to be even more successful when you are taking your test.**

All that we ask is that you email us your feedback about your study guide. Please let us know what you thought about it – whether that is good, bad or indifferent.

To get your **FREE Test Taking Tips DVD**, email freedvd@studyguideteam.com with “FREE DVD” in the subject line and the following information in the body of the email:

a. The title of your study guide.

b. Your product rating on a scale of 1-5, with 5 being the highest rating.

c. Your feedback about the study guide. What did you think of it?

d. Your full name and shipping address to send your free DVD.

If you have any questions or concerns, please don’t hesitate to contact us at freedvd@studyguideteam.com.

Thanks again!

CHSPE Preparation Book

CHSPE Study Guide with Practice Test Questions for the California High School Proficiency Exam [3rd Edition]

TPB Publishing

Written and edited by TPB Publishing.

ISBN 13: 9781628459265
ISBN 10: 1628459263

Table of Contents

Quick Overview

As you draw closer to taking your exam, effective preparation becomes more and more important. Thankfully, you have this study guide to help you get ready. Use this guide to help keep your studying on track and refer to it often.

This study guide contains several key sections that will help you be successful on your exam. The guide contains tips for what you should do the night before and the day of the test. Also included are test-taking tips. Knowing the right information is not always enough. Many well-prepared test takers struggle with exams. These tips will help equip you to accurately read, assess, and answer test questions.

A large part of the guide is devoted to showing you what content to expect on the exam and to helping you better understand that content. In this guide are practice test questions so that you can see how well you have grasped the content. Then, answer explanations are provided so that you can understand why you missed certain questions.

Don't try to cram the night before you take your exam. This is not a wise strategy for a few reasons. First, your retention of the information will be low. Your time would be better used by reviewing information you already know rather than trying to learn a lot of new information. Second, you will likely become stressed as you try to gain a large amount of knowledge in a short amount of time. Third, you will be depriving yourself of sleep. So be sure to go to bed at a reasonable time the night before. Being well-rested helps you focus and remain calm.

Be sure to eat a substantial breakfast the morning of the exam. If you are taking the exam in the afternoon, be sure to have a good lunch as well. Being hungry is distracting and can make it difficult to focus. You have hopefully spent lots of time preparing for the exam. Don't let an empty stomach get in the way of success!

When travelling to the testing center, leave earlier than needed. That way, you have a buffer in case you experience any delays. This will help you remain calm and will keep you from missing your appointment time at the testing center.

Be sure to pace yourself during the exam. Don't try to rush through the exam. There is no need to risk performing poorly on the exam just so you can leave the testing center early. Allow yourself to use all of the allotted time if needed.

Remain positive while taking the exam even if you feel like you are performing poorly. Thinking about the content you should have mastered will not help you perform better on the exam.

Once the exam is complete, take some time to relax. Even if you feel that you need to take the exam again, you will be well served by some down time before you begin studying again. It's often easier to convince yourself to study if you know that it will come with a reward!

Test-Taking Strategies

1. Predicting the Answer

When you feel confident in your preparation for a multiple-choice test, try predicting the answer before reading the answer choices. This is especially useful on questions that test objective factual knowledge. By predicting the answer before reading the available choices, you eliminate the possibility that you will be distracted or led astray by an incorrect answer choice. You will feel more confident in your selection if you read the question, predict the answer, and then find your prediction among the answer choices. After using this strategy, be sure to still read all of the answer choices carefully and completely. If you feel unprepared, you should not attempt to predict the answers. This would be a waste of time and an opportunity for your mind to wander in the wrong direction.

2. Reading the Whole Question

Too often, test takers scan a multiple-choice question, recognize a few familiar words, and immediately jump to the answer choices. Test authors are aware of this common impatience, and they will sometimes prey upon it. For instance, a test author might subtly turn the question into a negative, or he or she might redirect the focus of the question right at the end. The only way to avoid falling into these traps is to read the entirety of the question carefully before reading the answer choices.

3. Looking for Wrong Answers

Long and complicated multiple-choice questions can be intimidating. One way to simplify a difficult multiple-choice question is to eliminate all of the answer choices that are clearly wrong. In most sets of answers, there will be at least one selection that can be dismissed right away. If the test is administered on paper, the test taker could draw a line through it to indicate that it may be ignored; otherwise, the test taker will have to perform this operation mentally or on scratch paper. In either case, once the obviously incorrect answers have been eliminated, the remaining choices may be considered. Sometimes identifying the clearly wrong answers will give the test taker some information about the correct answer. For instance, if one of the remaining answer choices is a direct opposite of one of the eliminated answer choices, it may well be the correct answer. The opposite of obviously wrong is obviously right! Of course, this is not always the case. Some answers are obviously incorrect simply because they are irrelevant to the question being asked. Still, identifying and eliminating some incorrect answer choices is a good way to simplify a multiple-choice question.

4. Don't Overanalyze

Anxious test takers often overanalyze questions. When you are nervous, your brain will often run wild, causing you to make associations and discover clues that don't actually exist. If you feel that this may be a problem for you, do whatever you can to slow down during the test. Try taking a deep breath or counting to ten. As you read and consider the question, restrict yourself to the particular words used by the author. Avoid thought tangents about what the author *really* meant, or what he or she was *trying* to say. The only things that matter on a multiple-choice test are the words that are actually in the question. You must avoid reading too much into a multiple-choice question, or supposing that the writer meant something other than what he or she wrote.

5. No Need for Panic

It is wise to learn as many strategies as possible before taking a multiple-choice test, but it is likely that you will come across a few questions for which you simply don't know the answer. In this situation, avoid panicking. Because most multiple-choice tests include dozens of questions, the relative value of a single wrong answer is small. As much as possible, you should compartmentalize each question on a multiple-choice test. In other words, you should not allow your feelings about one question to affect your success on the others. When you find a question that you either don't understand or don't know how to answer, just take a deep breath and do your best. Read the entire question slowly and carefully. Try rephrasing the question a couple of different ways. Then, read all of the answer choices carefully. After eliminating obviously wrong answers, make a selection and move on to the next question.

6. Confusing Answer Choices

When working on a difficult multiple-choice question, there may be a tendency to focus on the answer choices that are the easiest to understand. Many people, whether consciously or not, gravitate to the answer choices that require the least concentration, knowledge, and memory. This is a mistake. When you come across an answer choice that is confusing, you should give it extra attention. A question might be confusing because you do not know the subject matter to which it refers. If this is the case, don't eliminate the answer before you have affirmatively settled on another. When you come across an answer choice of this type, set it aside as you look at the remaining choices. If you can confidently assert that one of the other choices is correct, you can leave the confusing answer aside. Otherwise, you will need to take a moment to try to better understand the confusing answer choice. Rephrasing is one way to tease out the sense of a confusing answer choice.

7. Your First Instinct

Many people struggle with multiple-choice tests because they overthink the questions. If you have studied sufficiently for the test, you should be prepared to trust your first instinct once you have carefully and completely read the question and all of the answer choices. There is a great deal of research suggesting that the mind can come to the correct conclusion very quickly once it has obtained all of the relevant information. At times, it may seem to you as if your intuition is working faster even than your reasoning mind. This may in fact be true. The knowledge you obtain while studying may be retrieved from your subconscious before you have a chance to work out the associations that support it. Verify your instinct by working out the reasons that it should be trusted.

8. Key Words

Many test takers struggle with multiple-choice questions because they have poor reading comprehension skills. Quickly reading and understanding a multiple-choice question requires a mixture of skill and experience. To help with this, try jotting down a few key words and phrases on a piece of scrap paper. Doing this concentrates the process of reading and forces the mind to weigh the relative importance of the question's parts. In selecting words and phrases to write down, the test taker thinks about the question more deeply and carefully. This is especially true for multiple-choice questions that are preceded by a long prompt.

9. Subtle Negatives

One of the oldest tricks in the multiple-choice test writer's book is to subtly reverse the meaning of a question with a word like *not* or *except*. If you are not paying attention to each word in the question, you can easily be led astray by this trick. For instance, a common question format is, "Which of the following is...?" Obviously, if the question instead is, "Which of the following is not...?," then the answer will be quite different. Even worse, the test makers are aware of the potential for this mistake and will include one answer choice that would be correct if the question were not negated or reversed. A test taker who misses the reversal will find what he or she believes to be a correct answer and will be so confident that he or she will fail to reread the question and discover the original error. The only way to avoid this is to practice a wide variety of multiple-choice questions and to pay close attention to each and every word.

10. Reading Every Answer Choice

It may seem obvious, but you should always read every one of the answer choices! Too many test takers fall into the habit of scanning the question and assuming that they understand the question because they recognize a few key words. From there, they pick the first answer choice that answers the question they believe they have read. Test takers who read all of the answer choices might discover that one of the latter answer choices is actually *more* correct. Moreover, reading all of the answer choices can remind you of facts related to the question that can help you arrive at the correct answer. Sometimes, a misstatement or incorrect detail in one of the latter answer choices will trigger your memory of the subject and will enable you to find the right answer. Failing to read all of the answer choices is like not reading all of the items on a restaurant menu: you might miss out on the perfect choice.

11. Spot the Hedges

One of the keys to success on multiple-choice tests is paying close attention to every word. This is never truer than with words like almost, most, some, and sometimes. These words are called "hedges" because they indicate that a statement is not totally true or not true in every place and time. An absolute statement will contain no hedges, but in many subjects, the answers are not always straightforward or absolute. There are always exceptions to the rules in these subjects. For this reason, you should favor those multiple-choice questions that contain hedging language. The presence of qualifying words indicates that the author is taking special care with his or her words, which is certainly important when composing the right answer. After all, there are many ways to be wrong, but there is only one way to be right! For this reason, it is wise to avoid answers that are absolute when taking a multiple-choice test. An absolute answer is one that says things are either all one way or all another. They often include words like *every*, *always*, *best*, and *never*. If you are taking a multiple-choice test in a subject that doesn't lend itself to absolute answers, be on your guard if you see any of these words.

12. Long Answers

In many subject areas, the answers are not simple. As already mentioned, the right answer often requires hedges. Another common feature of the answers to a complex or subjective question are qualifying clauses, which are groups of words that subtly modify the meaning of the sentence. If the question or answer choice describes a rule to which there are exceptions or the subject matter is complicated, ambiguous, or confusing, the correct answer will require many words in order to be expressed clearly and accurately. In essence, you should not be deterred by answer choices that seem excessively long. Oftentimes, the author of the text will not be able to write the correct answer without offering some qualifications and modifications. Your job is to read the answer choices thoroughly and completely and to select the one that most accurately and precisely answers the question.

13. Restating to Understand

Sometimes, a question on a multiple-choice test is difficult not because of what it asks but because of how it is written. If this is the case, restate the question or answer choice in different words. This process serves a couple of important purposes. First, it forces you to concentrate on the core of the question. In order to rephrase the question accurately, you have to understand it well. Rephrasing the question will concentrate your mind on the key words and ideas. Second, it will present the information to your mind in a fresh way. This process may trigger your memory and render some useful scrap of information picked up while studying.

14. True Statements

Sometimes an answer choice will be true in itself, but it does not answer the question. This is one of the main reasons why it is essential to read the question carefully and completely before proceeding to the answer choices. Too often, test takers skip ahead to the answer choices and look for true statements. Having found one of these, they are content to select it without reference to the question above. Obviously, this provides an easy way for test makers to play tricks. The savvy test taker will always read the entire question before turning to the answer choices. Then, having settled on a correct answer choice, he or she will refer to the original question and ensure that the selected answer is relevant. The mistake of choosing a correct-but-irrelevant answer choice is especially common on questions related to specific pieces of objective knowledge. A prepared test taker will have a wealth of factual knowledge at his or her disposal, and should not be careless in its application.

15. No Patterns

One of the more dangerous ideas that circulates about multiple-choice tests is that the correct answers tend to fall into patterns. These erroneous ideas range from a belief that B and C are the most common right answers, to the idea that an unprepared test-taker should answer "A-B-A-C-A-D-A-B-A." It cannot be emphasized enough that pattern-seeking of this type is exactly the WRONG way to approach a multiple-choice test. To begin with, it is highly unlikely that the test maker will plot the correct answers according to some predetermined pattern. The questions are scrambled and delivered in a random order. Furthermore, even if the test maker was following a pattern in the assignation of correct answers, there is no reason why the test taker would know which pattern he or she was using. Any attempt to discern a pattern in the answer choices is a waste of time and a distraction from the real work of taking the test. A test taker would be much better served by extra preparation before the test than by reliance on a pattern in the answers.

FREE DVD OFFER

Don't forget that doing well on your exam includes both understanding the test content and understanding how to use what you know to do well on the test. We offer a completely FREE Test Taking Tips DVD that covers world class test taking tips that you can use to be even more successful when you are taking your test.

All that we ask is that you email us your feedback about your study guide. To get your **FREE Test Taking Tips DVD**, email freedvd@studyguideteam.com with "FREE DVD" in the subject line and the following information in the body of the email:

- The title of your study guide.
- Your product rating on a scale of 1-5, with 5 being the highest rating.
- Your feedback about the study guide. What did you think of it?
- Your full name and shipping address to send your free DVD.

Introduction to the CHSPE

Function of the Test

The California High School Proficiency Examination is a voluntary test designed to provide students in the state of California with the legal equivalent of a high school diploma—a Certificate of Proficiency. Upon successfully passing the test, students may elect to exit high school early. The Certificate of Proficiency, granted by the California State Board of Education, is recognized as the equivalent of a high school diploma when individuals apply for federal civilian employment or federal financial aid in preparation to attend college.

Students taking this exam must be at least sixteen years old and enrolled in tenth grade for at least one academic year by the time the exam will be conducted. It is important to note that passing this exam is not the same as completing all high school coursework, and some colleges may not find it acceptable for admission.

Test Administration

The CHSPE test is typically offered three times during the calendar year: once in the fall, once in the spring, and once in the summer. The exam is offered at over sixty testing centers across the state of California. Students can register for the exam by visiting https://www.chspe.net/registration/form/. On the day of the exam, students can choose to take the CHSPE in its entirety or to simply take only the mathematics section, the language subtest, the reading subtest, or some combination of sections. However, the testing fee is the same, regardless of the student's choice.

There is no waiting period to retake the CHSPE exam, and there is no limit on the number of times an individual can retest. If an individual has not passed a particular subtest or section (reading subtest, language subtest, or mathematics section), he or she can simply retake that subtest or section only rather than the entire exam. However, if a student decides to retake the language subtest, he or she must complete both the multiple-choice questions and the writing task. It is important to note that a test taker's most recent test scores are the only ones that will be displayed on his or her official score report. Accommodations may be requested by completing an Accommodation Request Form during the registration process. In some instances, supporting documentation may be required. A student may be assigned to take the CHSPE exam at a different testing site if a particular location is not able to meet his or her permitted accommodations.

Test Format

The CHSPE test is comprised of two sections: English language arts and mathematics. The English language arts section is broken down into reading and language subtests. The reading subtest has 84 multiple-choice questions split between the two subject areas of comprehension and vocabulary. Comprehension is divided into specific content clusters (initial understanding, interpretation, critical analysis, and strategies), and vocabulary is also divided into specific content clusters (synonyms, multiple meaning words, and context clues). The language subtest has 48 multiple-choice questions divided between the two subject areas of mechanics and expressions. Mechanics is divided into specific content clusters (capitalization, usage, and punctuation), and expressions is also divided into specific content clusters (sentence structure, prewriting, and content and organization). The language subtest also contains a writing task that requires students to draft an essay explaining their opinion on an issue with reasoning to support that opinion. The mathematics section of the exam has 50 multiple-choice questions that are divided between specific content clusters (number sense and operations; patterns, relationships, and algebra; data, statistics, and probability; and geometry and measurement). Test takers are allowed to bring a basic calculator with them to the test site to use when completing the mathematics section. Students will have three and a half hours to complete the entire test.

CHSPE Test – English Language Arts	
Reading Subtest	
Subject Areas	*Questions (multiple-choice)*
Comprehension	54
Vocabulary	30
Total	84
Language Subtest	
Subject Areas	*Questions (multiple-choice)*
Mechanics	24
Expressions	24
Total	48
Writing Task	1
CHSPE Test – Mathematics	
	Questions (multiple-choice)
Total	50
182 Total Multiple-Choice Questions	3.5 Hours to Complete

Scoring

Individuals are not penalized for guessing when answering the multiple-choice questions. The range of scores for the multiple-choice questions within the reading and language subtests and the mathematics section is 250-450, and the range of scores for the writing task that is found within the language subtest is 1-5. The passing score for the reading subtest and the mathematics section of the CHSPE exam is 350. The passing score for the language subtest of the CHSPE is made up of a combination of the writing task and the multiple-choice questions, and is broken down as follows:

- If a student earns a score of 2 or less on the writing task, he or she cannot pass the language subtest regardless of how many multiple-choice questions were answered correctly.
- If a student earns a score of 2.5 on the writing task, he or she will need a score of 365 on the multiple-choice questions on this subtest in order to pass.
- If a student earns a score of 3 on the writing task, he or she will need a score of 350 on the multiple-choice questions on this subtest in order to pass.
- If a student earns a score of 3.5 or higher on the writing task, he or she will need a score of 342 on the multiple-choice questions on this subtest in order to pass.

Students will receive their test results about five weeks following completion of the CHSPE exam. Students who pass the exam will also receive their Certificate of Proficiency.

Study Prep Plan for the CHSPE

1. **Schedule** - Use one of our study schedules below or come up with one of your own.

2. **Relax** - Test anxiety can hurt event the best students. There are many ways to reduce stress. Find the one that works best for you.

3. **Execute** - Once you have a good plan in place, be sure to stick to it.

Sample Study Plans

One Week Study Schedule

Day 1	Introduction to the CHSPE
Day 2	Reading
Day 3	Language
Day 4	Writing
Day 5	Math
Day 6	Practice Questions
Day 7	Take Your Exam!

Two Week Study Schedule

Day 1	Introduction to the CHSPE	Day 8	Patterns, Relationships, and Algebra
Day 2	Comprehension	Day 9	Data, Statistics, and Probability
Day 3	Vocabulary	Day 10	Geometry and Measurement
Day 4	Mechanics	Day 11	Practice Questions
Day 5	Expressions	Day 12	Review Answer Explanations
Day 6	Writing	Day 13	(Study Break)
Day 7	Number Sense and Operations	Day 14	Take Your Exam!

One Month Study Schedule

Day 1	Introduction to the CHSPE	Day 11	Punctuation	Day 21	Properties of Exponents
Day 2	Initial Understanding	Day 12	Sentence Structure	Day 22	Positive Rational Roots
Day 3	Interpretation	Day 13	Prewriting	Day 23	Scientific Notation
Day 4	Critical Analysis and Strategies	Day 14	Content and Organization	Day 24	Patterns, Relationships, and Algebra
Day 5	Synonyms	Day 15	Practice Questions	Day 25	Data, Statistics, and Probability
Day 6	Multiple Meaning Words	Day 16	Review Answer Explanations	Day 26	Geometry and Measurement
Day 7	Context Clues	Day 17	Writing Task	Day 27	Practice Questions
Day 8	Practice Questions	Day 18	Addition and Subtraction	Day 28	Review Answer Explanations
Day 9	Review Answer Explanations	Day 19	Multiplication and Division	Day 29	(Study Break)
Day 10	Usage and Capitalization	Day 20	Percent Problems	Day 30	Take Your Exam!

Reading

Comprehension

Initial Understanding

The Purpose of a Passage

No matter the genre or format, all authors are writing to persuade, inform, entertain, or express feelings. Often, these purposes are blended, with one dominating the rest. It's useful to learn to recognize the author's intent.

Persuasive writing is used to persuade or convince readers of something. It often contains two elements: the argument and the counterargument. The argument takes a stance on an issue, while the counterargument pokes holes in the opposition's stance. Authors rely on logic, emotion, and writer credibility to persuade readers to agree with them. If readers are opposed to the stance before reading, they are unlikely to adopt that stance. However, those who are undecided or committed to the same stance are more likely to agree with the author.

Informative writing tries to teach or inform. Workplace manuals, instructor lessons, statistical reports and cookbooks are examples of informative texts. Informative writing is usually based on facts and is often void of emotion and persuasion. Informative texts generally contain statistics, charts, and graphs. Though most informative texts lack a persuasive agenda, readers must examine the text carefully to determine whether one exists within a given passage.

Stories or narratives are designed to entertain. When you go to the movies, you often want to escape for a few hours, not necessarily to think critically. Entertaining writing is designed to delight and engage the reader. However, sometimes this type of writing can be woven into more serious materials, such as persuasive or informative writing to hook the reader before transitioning into a more scholarly discussion.

Emotional writing works to evoke the reader's feelings, such as anger, euphoria, or sadness. The connection between reader and author is an attempt to cause the reader to share the author's intended emotion or tone. Sometimes in order to make a piece more poignant, the author simply wants readers to feel the same emotions that the author has felt. Other times, the author attempts to persuade or manipulate the reader into adopting his stance. While it's okay to sympathize with the author, be aware of the individual's underlying intent.

Types of Passages

Writing can be classified under four passage types: narrative, expository, descriptive (sometimes called technical), and persuasive. Though these types are not mutually exclusive, one form tends to dominate the rest. By recognizing the *type* of passage you're reading, you gain insight into *how* you should read. If you're reading a narrative, you can assume the author intends to entertain, which means you may skim the text without losing meaning. A technical document might require a close read because skimming the passage might cause the reader to miss salient details.

1. *Narrative* writing, at its core, is the art of storytelling. For a narrative to exist, certain elements must be present. First, it must have characters. While many characters are human, characters could be defined as anything that thinks, acts, and talks like a human. For example, many recent movies, such as

Lord of the Rings and *The Chronicles of Narnia*, include animals, fantastical creatures, and even trees that behave like humans. Second, it must have a plot or sequence of events. Typically, those events follow a standard plot diagram, but recent trends start *in medias res* or in the middle (near the climax). In this instance, foreshadowing and flashbacks often fill in plot details. Finally, along with characters and a plot, there must also be conflict. Conflict is usually divided into two types: internal and external. Internal conflict indicates the character is in turmoil and is presented through the character's thoughts. External conflicts are visible. Types of external conflict include a person versus nature, another person, or society.

2. *Expository writing is detached and to the point. Since expository* writing is designed to instruct or inform, it usually involves directions and steps written in second person ("you" voice) and lacks any persuasive or narrative elements. Sequence words such as *first*, *second*, and *third*, or *in the first place*, *secondly*, and *lastly* are often given to add fluency and cohesion. Common examples of expository writing include instructor's lessons, cookbook recipes, and repair manuals.

3. Due to its empirical nature, *technical* writing is filled with steps, charts, graphs, data, and statistics. The goal of technical writing is to advance understanding in a field through the scientific method. Experts such as teachers, doctors, or mechanics use words unique to the profession in which they operate. These words, which often incorporate acronyms, are called *jargon*. Technical writing is a type of expository writing but is not meant to be understood by the general public. Instead, technical writers assume readers have received a formal education in a particular field of study and need no explanation as to what the jargon means. Imagine a doctor trying to understand a diagnostic reading for a car or a mechanic trying to interpret lab results. Only professionals with proper training will fully comprehend the text.

4. *Persuasive* writing is designed to change opinions and attitudes. The topic, stance, and arguments are found in the thesis, positioned near the end of the introduction. Later supporting paragraphs offer relevant quotations, paraphrases, and summaries from primary or secondary sources, which are then interpreted, analyzed, and evaluated. The goal of persuasive writers is not to stack quotes but to develop original ideas by using sources as a starting point. Good persuasive writing makes powerful arguments with valid sources and thoughtful analysis. Poor persuasive writing is riddled with bias and logical fallacies. Sometimes logical and illogical arguments are sandwiched together in the same piece. Therefore, readers should display skepticism when reading persuasive arguments.

Text Structure

Depending on what the author is attempting to accomplish, certain formats or text structures work better than others. For example, a sequence structure might work for narration but not when identifying similarities and differences between dissimilar concepts. Similarly, a comparison-contrast structure is not useful for narration. It's the author's job to put the right information in the correct format.

Readers should be familiar with the five main literary structures:

1. *Sequence* structure (sometimes referred to as the order structure) is when the order of events proceed in a predictable order. In many cases, this means the text goes through the plot elements: exposition, rising action, climax, falling action, and resolution. Readers are introduced to characters, setting, and conflict in the exposition. In the rising action, there's an increase in tension and suspense. The climax is the height of tension and the point of no return. Tension decreases during the falling

action. In the resolution, any conflicts presented in the exposition are solved, and the story concludes. An informative text that is structured sequentially will often go in order from one step to the next.

2. In the *problem-solution* structure, authors identify a potential problem and suggest a solution. This form of writing is usually divided into two paragraphs and can be found in informational texts. For example, cell phone, cable, and satellite providers use this structure in manuals to help customers troubleshoot or identify problems with services or products.

3. When authors want to discuss similarities and differences between separate concepts, they arrange thoughts in a *comparison-contrast* paragraph structure. Venn diagrams are an effective graphic organizer for comparison-contrast structures because they feature two overlapping circles that can be used to organize similarities and differences. A comparison-contrast essay organizes one paragraph based on similarities and another based on differences. A comparison-contrast essay can also be arranged with the similarities and differences of individual traits addressed within individual paragraphs. Words such as *however*, *but*, and *nevertheless* help signal a contrast in ideas.

4. *Descriptive* writing structure is designed to appeal to your senses. Much like an artist who constructs a painting, good descriptive writing builds an image in the reader's mind by appealing to the five senses: sight, hearing, taste, touch, and smell. However, overly descriptive writing can become tedious; whereas sparse descriptions can make settings and characters seem flat. Good authors strike a balance by applying descriptions only to passages, characters, and settings that are integral to the plot.

5. Passages that use the *cause and effect* structure are simply asking *why* by demonstrating some type of connection between ideas. Words such as *if*, *since*, *because*, *then*, or *consequently* indicate relationship. By switching the order of a complex sentence, the writer can rearrange the emphasis on different clauses. Saying *If Sheryl is late, we'll miss the dance* is different from saying *We'll miss the dance if Sheryl is late*. One emphasizes Sheryl's tardiness while the other emphasizes missing the dance. Paragraphs can also be arranged in a cause and effect format. Since the format—before and after—is sequential, it is useful when authors wish to discuss the impact of choices. Researchers often apply this paragraph structure to the scientific method.

Point of View

Point of view is an important writing device to consider. In fiction writing, point of view refers to who tells the story or from whose perspective readers are observing the story. In non-fiction writing, the *point of view* refers to whether the author refers to himself/herself, his/her readers, or chooses not to mention either. Whether fiction or nonfiction, the author will carefully consider the impact the perspective will have on the purpose and main point of the writing.

First-person point of view: The story is told from the writer's perspective. In fiction, this would mean that the main character is also the narrator. First-person point of view is easily recognized by the use of personal pronouns such as *I*, *me*, *we*, *us*, *our*, *my*, and *myself*.

Third-person point of view: In a more formal essay, this would be an appropriate perspective because the focus should be on the subject matter, not the writer or the reader. Third-person point of view is recognized by the use of the pronouns *he*, *she*, *they*, and *it*.

In fiction writing, third-person point of view has a few variations.

- *Third-person limited* point of view refers to a story told by a narrator who has access to the thoughts and feelings of just one character.
- In *third-person omniscient* point of view, the narrator has access to the thoughts and feelings of all the characters.
- In *third-person objective* point of view, the narrator is like a fly on the wall and can see and hear what the characters do and say but does not have access to their thoughts and feelings.

Second-person point of view: This point of view isn't commonly used in fiction or nonfiction writing because it directly addresses the reader using the pronouns *you*, *your*, and *yourself*. Second-person perspective is more appropriate in direct communication, such as business letters or emails.

Point of View	Pronouns Used
First person	I, me, we, us, our, my, myself
Second person	You, your, yourself
Third person	He, she, it, they

Main Ideas and Supporting Details

Topics and main ideas are critical parts of writing. The **topic** is the subject matter of the piece. An example of a topic would be *the use of cell phones in a classroom*.

The **main idea** is what the writer wants to say about that topic. A writer may make the point that the use of cell phones in a classroom is a serious problem that must be addressed in order for students to learn better. Therefore, the topic is cell phone usage in a classroom, and the main idea is that it's *a serious problem needing to be addressed*. The topic can be expressed in a word or two, but the main idea should be a complete thought.

An author will likely identify the topic immediately within the title or the first sentence of the passage. The main idea is usually presented in the introduction. In a single passage, the main idea may be identified in the first or last sentence, but it will most likely be directly stated and easily recognized by the reader. Because it is not always stated immediately in a passage, it's important that readers carefully read the entire passage to identify the main idea.

The main idea should not be confused with the thesis statement. A **thesis statement** is a clear statement of the writer's specific stance and can often be found in the introduction of a nonfiction piece. The thesis is a specific sentence (or two) that offers the direction and focus of the discussion.

In order to illustrate the main idea, a writer will use **supporting details**, which provide evidence or examples to help make a point. Supporting details are typically found in nonfiction pieces that seek to inform or persuade the reader.

In the example of cell phone usage in the classroom, where the author's main idea is to show the seriousness of this problem and the need to "unplug", supporting details would be critical for effectively making that point. Supporting details used here might include statistics on a decline in student focus and studies showing the impact of digital technology usage on students' attention spans. The author could also include testimonies from teachers surveyed on the topic.

It's important that readers evaluate the author's supporting details to be sure that they are credible, provide evidence of the author's point, and directly support the main idea. Although shocking statistics grab readers' attention, their use may provide ineffective information in the piece. Details like this are crucial to understanding the passage and evaluating how well the author presents his or her argument and evidence.

Also remember that when most authors write, they want to make a point or send a message. This point or message of a text is known as the theme. Authors may state themes explicitly, like in *Aesop's Fables*. More often, especially in modern literature, readers must infer the theme based on text details. Usually after carefully reading and analyzing an entire text, the theme emerges. Typically, the longer the piece, the more themes you will encounter, though often one theme dominates the rest, as evidenced by the author's purposeful revisiting of it throughout the passage.

Interpretation

Evaluating a Passage

Determining conclusions requires being an active reader, as a reader must make a prediction and analyze facts to identify a conclusion. There are a few ways to determine a logical conclusion, but careful reading is the most important. It's helpful to read a passage a few times, noting details that seem important to the piece. A reader should also identify key words in a passage to determine the logical conclusion or determination that flows from the information presented.

Textual evidence within the details helps readers draw a conclusion about a passage. *Textual evidence* refers to information—facts and examples that support the main point. Textual evidence will likely come from outside sources and can be in the form of quoted or paraphrased material. In order to draw a conclusion from evidence, it's important to examine the credibility and validity of that evidence as well as how (and if) it relates to the main idea.

If an author presents a differing opinion or a *counter-argument* in order to refute it, the reader should consider how and why this information is being presented. It is meant to strengthen the original argument and shouldn't be confused with the author's intended conclusion, but it should also be considered in the reader's final evaluation.

Sometimes, authors explicitly state the conclusion they want readers to understand. Alternatively, a conclusion may not be directly stated. In that case, readers must rely on the implications to form a logical conclusion:

> On the way to the bus stop, Michael realized his homework wasn't in his backpack. He ran back to the house to get it and made it back to the bus just in time.

In this example, though it's never explicitly stated, it can be inferred that Michael is a student on his way to school in the morning. When forming a conclusion from implied information, it's important to read the text carefully to find several pieces of evidence to support the conclusion.

Summarizing is an effective way to draw a conclusion from a passage. A summary is a shortened version of the original text, written by the reader in his/her own words. Focusing on the main points of the original text and including only the relevant details can help readers reach a conclusion. It's important to retain the original meaning of the passage.

Like summarizing, *paraphrasing* can also help a reader fully understand different parts of a text. Paraphrasing calls for the reader to take a small part of the passage and list or describe its main points. Paraphrasing is more than rewording the original passage, though. It should be written in the reader's own words, while still retaining the meaning of the original source. This will indicate an understanding of the original source, yet still help the reader expand on his/her interpretation.

Readers should pay attention to the *sequence*, or the order in which details are laid out in the text, as this can be important to understanding its meaning as a whole. Writers will often use transitional words to help the reader understand the order of events and to stay on track. Words like *next, then, after*, and *finally* show that the order of events is important to the author. In some cases, the author omits these transitional words, and the sequence is implied. Authors may even purposely present the information out of order to make an impact or have an effect on the reader. An example might be when a narrative writer uses *flashback* to reveal information.

Responding to a Passage

There are a few ways for readers to engage actively with the text, such as making inferences and predictions. An *inference* refers to a point that is implied (as opposed to directly-stated) by the evidence presented:

> Bradley packed up all of the items from his desk in a box and said goodbye to his coworkers for the last time.

From this sentence, though it is not directly stated, readers can infer that Bradley is leaving his job. It's necessary to use inference in order to draw conclusions about the meaning of a passage. Authors make implications through character dialogue, thoughts, effects on others, actions, and looks. Like in life, readers must assemble all the clues to form a complete picture.

When making an inference about a passage, it's important to rely only on the information that is provided in the text itself. This helps readers ensure that their conclusions are valid.

Readers will also find themselves making predictions when reading a passage or paragraph. *Predictions* are guesses about what's going to happen next. Readers can use prior knowledge to help make accurate predictions. Prior knowledge is best utilized when readers make links between the current text, previously read texts, and life experiences. Some texts use suspense and foreshadowing to captivate readers:

A cat darted across the street just as the car came careening around the curve.

One unfortunate prediction might be that the car will hit the cat. Of course, predictions aren't always accurate, so it's important to read carefully to the end of the text to determine the accuracy of predictions.

Critical Analysis

It's important to read any piece of writing critically. The goal is to discover the point and purpose of what the author is writing about through analysis. It's also crucial to establish the point or stance the author has taken on the topic of the piece. After determining the author's perspective, readers can then more effectively develop their own viewpoints on the subject of the piece.

It is important to distinguish between *fact and opinion* when reading a piece of writing. A fact is information that can be proven true. If information can be disproved, it is not a fact. For example, water freezes at or below thirty-two degrees Fahrenheit. An argument stating that water freezes at seventy degrees Fahrenheit cannot be supported by data and is therefore not a fact. Facts tend to be associated with science, mathematics, and statistics. Opinions are information open to debate. Opinions are often tied to subjective concepts like equality, morals, and rights. They can also be controversial.

Authors often use words like *think, feel, believe,* or *in my opinion* when expressing opinion, but these words won't always appear in an opinion piece, especially if it is formally written. An author's opinion may be backed up by facts, which gives it more credibility, but that opinion should not be taken as fact. A critical reader should be suspect of an author's opinion, especially if it is only supported by other opinions.

Fact	**Opinion**
There are 9 innings in a game of baseball.	Baseball games run too long.
James Garfield was assassinated on July 2, 1881.	James Garfield was a good president.
McDonalds has stores in 118 countries.	McDonalds has the best hamburgers.

Critical readers examine the facts used to support an author's argument. They check the facts against other sources to be sure those facts are correct. They also check the validity of the sources used to be sure those sources are credible, academic, and/or peer- reviewed. Consider that when an author uses another person's opinion to support his or her argument, even if it is an expert's opinion, it is still only an opinion and should not be taken as fact. A strong argument uses valid, measurable facts to support ideas. Even then, the reader may disagree with the argument as it may be rooted in his or her personal beliefs.

An authoritative argument may use the facts to sway the reader. Because of this, a writer may choose to only use the information and expert opinion that supports his or her viewpoint.

If the argument is that wind energy is the best solution, the author will use facts that support this idea. That same author may leave out relevant facts on solar energy. The way the author uses facts can influence the reader, so it's important to consider the facts being used, how those facts are being presented, and what information might be left out.

Critical readers should also look for errors in the argument such as logical fallacies and bias. A *logical fallacy* is a flaw in the logic used to make the argument. Logical fallacies include slippery slope, straw man, and begging the question. Authors can also reflect *bias* if they ignore an opposing viewpoint or present their side in an unbalanced way. A strong argument considers the opposition and finds a way to refute it. Critical readers should look for an unfair or one-sided presentation of the argument and be skeptical, as a bias may be present. Even if this bias is unintentional, if it exists in the writing, the reader should be wary of the validity of the argument.

Readers should also look for the use of *stereotypes,* which refer to specific groups. Stereotypes are often negative connotations about a person or place and should always be avoided. When a critical reader finds stereotypes in a piece of writing, they should immediately be critical of the argument and consider the validity of anything the author presents. Stereotypes reveal a flaw in the writer's thinking and may suggest a lack of knowledge or understanding about the subject.

Strategies

Predictions

Some texts use suspense and foreshadowing to captivate readers. For example, an intriguing aspect of murder mysteries is that the reader is never sure of the culprit until the author reveals the individual's identity. Authors often build suspense and add depth and meaning to a work by leaving clues to provide hints or predict future events in the story; this is called foreshadowing. While some instances of foreshadowing are subtle, others are quite obvious.

Inferences

Another way to read actively is to identify examples of inference within text. Making an inference requires the reader to read between the lines and look for what is *implied* rather than what is directly stated. That is, using information that is known from the text, the reader is able to make a logical assumption about information that is *not* directly stated but is probably true.

Authors employ literary devices such as tone, characterization, and theme to engage the audience by showing details of the story instead of merely telling them. For example, if an author said *Bob is selfish*, there's little left to infer. If the author said, *Bob cheated on his test, ignored his mom's calls, and parked illegally*, the reader can infer Bob is selfish. Authors also make implications through character dialogue, thoughts, effects on others, actions, and looks. Like in life, readers must assemble all the clues to form a complete picture.

Read the following passage:

> "Hey, do you wanna meet my new puppy?" Jonathan asked.
>
> "Oh, I'm sorry but please don't—" Jacinta began to protest, but before she could finish, Jonathan had already opened the passenger side door of his car and a perfect white ball of fur came bouncing towards Jacinta.
>
> "Isn't he the cutest?" beamed Jonathan.
>
> "Yes—achoo!—he's pretty—aaaachooo!!—adora—aaa—aaaachoo!" Jacinta managed to say in between sneezes. "But if you don't mind, I—I—achoo!—need to go inside."

Which of the following can be inferred from Jacinta's reaction to the puppy?

a. she hates animals
b. she is allergic to dogs
c. she prefers cats to dogs
d. she is angry at Jonathan

An inference requires the reader to consider the information presented and then form their own idea about what is probably true. Based on the details in the passage, what is the best answer to the question? Important details to pay attention to include the tone of Jacinta's dialogue, which is overall polite and apologetic, as well as her reaction itself, which is a long string of sneezes. Answer choices (a) and (d) both express strong emotions ("hates" and "angry") that are not evident in Jacinta's speech or actions. Answer choice (c) mentions cats, but there is nothing in the passage to indicate Jacinta's feelings about cats. Answer choice (b), "she is allergic to dogs," is the most logical choice—based on the fact that she began sneezing as soon as a fluffy dog approached her, it makes sense to guess that Jacinta might be allergic to dogs. So even though Jacinta never directly states, "Sorry, I'm allergic to dogs!" using the clues in the passage, it is still reasonable to guess that this is true.

Making inferences is crucial for readers of literature because literary texts often avoid presenting complete and direct information to readers about characters' thoughts or feelings, or they present this information in an unclear way, leaving it up to the reader to interpret clues given in the text. In order to make inferences while reading, readers should ask themselves:

- What details are being presented in the text?
- Is there any important information that seems to be missing?
- Based on the information that the author *does* include, what else is probably true?
- Is this inference reasonable based on what is already known?

Conclusions

Active readers should also draw conclusions. When doing so, the reader should ask the following questions: What is this piece about? What does the author believe? Does this piece have merit? Do I believe the author? Would this piece support my argument? The reader should first determine the author's intent. Identify the author's viewpoint and connect relevant evidence to support it. Readers may then move to the most important step: deciding whether to agree and determining whether they are correct. Always read cautiously and critically. Interact with text, and record reactions in the margins. These active reading skills help determine not only what the author thinks, but what you as the reader thinks.

Vocabulary

Synonyms

Synonyms are words that mean the same or nearly the same thing in the same language. When presented with several words and asked to choose the synonym, more than one word may be similar to the original. However, one word is generally the strongest match. Synonyms should always share the same part of speech. For instance, *shy* and *timid* are both adjectives and hold similar meanings. The words *shy* and *loner* are similar, but *shy* is an adjective while *loner* is a noun. Another way to test for the best synonym is to reread the sentence with each possible word and determine which one makes the most sense. Consider the following sentence: *He will love you forever.*

Now consider the words: adore, sweet, kind, and nice. They seem similar, but when used in the following applications with the initial sentence, not all of them are synonyms for *love.*

> He will *adore* you forever.
>
> He will *sweet* you forever.
>
> He will *kind* you forever.
>
> He will *nice* you forever.

In the first sentence, the word *love* is used as a verb. The best synonym from the list that shares the same part of speech is *adore. Adore* is a verb, and when substituted in the sentence, it is the only substitution that makes grammatical and semantic sense.

Synonyms can be found for nouns, adjectives, verbs, adverbs, and prepositions. Here are some examples of synonyms from different parts of speech:

- Nouns: clothes, wardrobe, attire, apparel
- Verbs: run, sprint, dash
- Adjectives: fast, quick, rapid, swift
- Adverbs: slowly, nonchalantly, leisurely
- Prepositions: near, proximal, neighboring, close

Here are several more examples of synonyms in the English language:

Word	Synonym	Meaning
smart	intelligent	having or showing a high level of intelligence
exact	specific	clearly identified
almost	nearly	not quite but very close
to annoy	to bother	to irritate
to answer	to reply	to form a written or verbal response
building	edifice	a structure that stands on its own with a roof and four walls
business	commerce	the act of purchasing, negotiating, trading, and selling
defective	faulty	when a device is not working or not working well

Multiple Meaning Words

Homonyms

Homonyms are words that sound alike but carry different meanings. There are two different types of homonyms: homophones and homographs.

Homophones

Homophones are words that sound alike, but carry different meanings and spellings. In the English language, there are several examples of homophones. Consider the following list:

Word	Meaning	Homophone	Meaning
I'll	I + will	aisle	a specific lane between seats
allowed	past tense of the verb, 'to allow'	aloud	to utter a sound out loud
eye	a part of the body used for seeing	I	first-person singular
ate	the past tense of the verb, 'to eat'	eight	the number preceding the number nine
peace	the opposite of war	piece	part of a whole
seas	large bodies of natural water	seize	to take ahold of/to capture

Homographs

Homographs are words that share the same spelling but carry different meanings and different pronunciations. Consider the following list:

Word	Meaning	Homograph	Meaning
bass	fish	bass	musical instrument
bow	a weapon used to fire arrows	bow	to bend
Polish	of or from Poland	polish	a type of shine (n); to shine (v)
desert	dry, arid land	desert	to abandon

Context Clues

Familiarity with common prefixes, suffixes, and root words assists tremendously in unraveling the meaning of an unfamiliar word and making an educated guess as to its meaning. However, some words do not contain many easily-identifiable clues that point to their meaning. In this case, rather than looking at the elements within the word, it is useful to consider elements around the word—i.e., its context. *Context* refers to the other words and information within the sentence or surrounding sentences that indicate the unknown word's probable meaning. The following sentences provide context for the potentially-unfamiliar word *quixotic*:

> Rebecca had never been one to settle into a predictable, ordinary life. Her quixotic personality led her to leave behind a job with a prestigious law firm in Manhattan and move halfway around the world to pursue her dream of becoming a sushi chef in Tokyo.

A reader unfamiliar with the word *quixotic* doesn't have many clues to use in terms of affixes or root meaning. The suffix *–ic* indicates that the word is an adjective, but that is it. In this case, then, a reader would need to look at surrounding information to obtain some clues about the word. Other adjectives in the passage include *predictable* and *ordinary*, things that Rebecca was definitely not, as indicated by "Rebecca had never been one to settle." Thus, a first clue might be that *quixotic* means the opposite of predictable.

The second sentence doesn't offer any other modifier of *personality* other than *quixotic*, but it does include a story that reveals further information about her personality. She had a stable, respectable job, but she decided to give it up to follow her dream. Combining these two ideas together, then—unpredictable and dream-seeking—gives the reader a general idea of what *quixotic* probably means. In fact, the root of the word is the character Don Quixote, a romantic dreamer who goes on an impulsive adventure.

While context clues are useful for making an approximate definition for newly-encountered words, these types of clues also come in handy when encountering common words that have multiple meanings. The word *reservation* is used differently in each the following sentences:

- That restaurant is booked solid for the next month; it's impossible to make a reservation unless you know somebody.
- The hospital plans to open a branch office inside the reservation to better serve Native American patients who cannot easily travel to the main hospital fifty miles away.
- Janet Clark is a dependable, knowledgeable worker, and I recommend her for the position of team leader without reservation.

All three sentences use the word to express different meanings. In fact, most words in English have more than one meaning—sometimes meanings that are completely different from one another. Thus, context can provide clues as to which meaning is appropriate in a given situation. A quick search in the dictionary reveals several possible meanings for *reservation*:

- An exception or qualification
- A tract of public land set aside, such as for the use of American Indian tribes
- An arrangement for accommodations, such as in a hotel, on a plane, or at a restaurant

Sentence A mentions a restaurant, making the third definition the correct one in this case. In sentence B, some context clues include Native Americans, as well as the implication that a reservation is a place—"inside the reservation," both of which indicate that the second definition should be used here. Finally, sentence C uses *without reservation* to mean "completely" or "without exception," so the first definition can be applied here.

Using context clues in this way can be especially useful for words that have multiple, widely varying meanings. If a word has more than one definition and two of those definitions are the opposite of each other, it is known as an *auto-antonym*—a word that can also be its own antonym. In the case of auto-antonyms, context clues are crucial to determine which definition to employ in a given sentence. For example, the word *sanction* can either mean "to approve or allow" or "a penalty." Approving and penalizing have opposite meanings, so *sanction* is an example of an auto-antonym. The following sentences reflect the distinction in meaning:

- In response to North Korea's latest nuclear weapons test, world leaders have called for harsher sanctions to punish the country for its actions.
- The general has sanctioned a withdrawal of troops from the area.

A context clue can be found in sentence A, which mentions "to punish." A punishment is similar to a penalty, so sentence A is using the word *sanction* according to this definition.

Other examples of auto-antonyms include *oversight*—"to supervise something" or "a missed detail"), *resign*—"to quit" or "to sign again, as a contract," and *screen*—"to show" or "to conceal." For these types of words, recognizing context clues is an important way to avoid misinterpreting the sentence's meaning.

Practice Questions

Directions for questions 1 – 35: Choose the word that means the same thing as the underlined word.

1. "I promise, I did not <u>instigate</u> the fight."
   a. begin
   b. ponder
   c. swim
   d. overwhelm

2. <u>Fortify</u> most nearly means
   a. foster
   b. barricade
   c. strengthen
   d. undermine

3. My girlfriend said, "Either you marry me, or I'm leaving you!" Not a very pleasant <u>ultimatum</u>.
   a. invitation
   b. journey
   c. greeting
   d. threat

4. <u>Wave</u> most nearly means
   a. flourish
   b. sink
   c. tide
   d. stagnant

5. Space is often referred to as the great <u>void</u>.
   a. around
   b. age
   c. empty
   d. diffuse

6. <u>Query</u> most nearly means
   a. bury
   b. wander
   c. ask
   d. praise

7. The thievery merited <u>severe</u> punishment.
   a. foul
   b. extreme
   c. cut
   d. shake

8. <u>Vexation</u> most nearly means
   a. boring
   b. pain
   c. revolving
   d. anger

9. The toddler, who had just learned to speak, seemed rather **loquacious**.
   a. verbose
   b. humorous
   c. silent
   d. cranky

10. <u>Grim</u> most nearly means
   a. frank
   b. dire
   c. corpse
   d. sharp

11. The bullies <u>disparaged</u> the younger boy, causing him to feel worthless.
   a. despaired
   b. belittled
   c. broke
   d. sparred

12. <u>Altitude</u> most nearly means
   a. behavior
   b. outlook
   c. haughtiness
   d. height

13. Though the valedictorian was very smart, he was too <u>egotistic</u> to have very many friends.
   a. conceited
   b. altruistic
   c. agrarian
   d. unconcerned

14. <u>Understand</u> most nearly means
   a. build
   b. fathom
   c. endow
   d. guilt

15. The car was jostled by the rocky <u>terrain</u>.
   a. firmament
   b. celestial
   c. arboreal
   d. ground

16. Mitigate most nearly means
   a. alleviate
   b. focus
   c. fiery
   d. conspire

17. The king's army was able to easily encompass his brother's army, and quickly crushed the rebellion.
   a. retreat
   b. brazen
   c. surround
   d. avoid

18. Archetype most nearly means
   a. ancestor
   b. literature
   c. policy
   d. model

19. After sleeping in his car for the past few months, the tiny hotel room seemed like a mansion by contrast.
   a. partisan
   b. variation
   c. revive
   d. improve

20. Irate most nearly means
   a. amiable
   b. innocent
   c. incensed
   d. tangible

21. Given his rude comments, her reaction seemed plausible.
   a. conceivable
   b. skeptical
   c. diurnal
   d. erudite

22. Discourse most nearly means
   a. praise
   b. disagreement
   c. speech
   d. path

23. The young boy seemed to be capable of nothing, other than his uncanny ability to exasperate his babysitter.
   a. annoy
   b. breathe
   c. disappoint
   d. stifle

24. Bellicose most nearly means
   a. loud
   b. angry
   c. pugnacious
   d. patriotic

25. Given the horrendous situation, the man's equanimity was appalling.
   a. justice
   b. hostility
   c. equine
   d. composure

26. Preemptive most nearly means
   a. beforehand
   b. prepare
   c. hidden
   d. initial

27. The rebels fought in order to liberate their brothers from the evil dictator.
   a. agitate
   b. instigate
   c. fracture
   d. release

28. Mediocre most nearly means
   a. excellent
   b. average
   c. intrusive
   d. inspiring

29. The jury was not impressed by the lazy defendant's moot argument.
   a. factual
   b. historical
   c. debatable
   d. exemplify

30. Bestow most nearly means
   a. take
   b. bequeath
   c. energize
   d. study

31. The starving man was disheartened when he reached the summit of the hill and realized that only a barren wasteland awaited him.
   a. fruitful
   b. infertile
   c. sumptuous
   d. lavish

32. <u>Refurbish</u> most nearly means
    a. renovate
    b. enlighten
    c. craven
    d. burnish

33. When she saw the crayon drawings on the wall, the mother had no choice but to <u>chastise</u> her sons.
    a. honor
    b. locate
    c. choose
    d. rebuke

34. <u>Sustain</u> most nearly means
    a. strength
    b. invigorate
    c. embolden
    d. expedite

35. When the boy saw how sincere the girl's apology was, he decided to <u>acquit</u> her of her faults.
    a. forgive
    b. stall
    c. acquire
    d. quit

Questions 36 – 39 are based on the following passage.

Smoking is Terrible

Smoking tobacco products is terribly destructive. A single cigarette contains over 4,000 chemicals, including 43 known carcinogens and 400 deadly toxins. Some of the most dangerous ingredients include tar, carbon monoxide, formaldehyde, ammonia, arsenic, and DDT. Smoking can cause numerous types of cancer including throat, mouth, nasal cavity, esophageal, gastric, pancreatic, renal, bladder, and cervical cancer.

Cigarettes contain a drug called nicotine, one of the most addictive substances known to man. Addiction is defined as a compulsion to seek the substance despite negative consequences. According to the National Institute of Drug Abuse, nearly 35 million smokers expressed a desire to quit smoking in 2015; however, more than 85 percent of those who struggle with addiction will not achieve their goal. Almost all smokers regret picking up that first cigarette. You would be wise to learn from their mistake if you have not yet started smoking.

According to the U.S. Department of Health and Human Services, 16 million people in the United States presently suffer from a smoking-related condition and nearly nine million suffer from a serious smoking-related illness. According to the Centers for Disease Control and Prevention (CDC), tobacco products cause nearly six million deaths per year. This number is projected to rise to over eight million deaths by 2030. Smokers, on average, die ten years earlier than their nonsmoking peers.

In the United States, local, state, and federal governments typically tax tobacco products, which leads to high prices. Nicotine users who struggle with addiction sometimes pay more for a pack of cigarettes than for a few gallons of gas. Additionally, smokers tend to stink. The smell of smoke is all-consuming and creates a pervasive nastiness. Smokers also risk staining their teeth and fingers with yellow residue from the tar.

Smoking is deadly, expensive, and socially unappealing. Clearly, smoking is not worth the risks.

36. Which of the following best describes the passage?
   a. Narrative
   b. Persuasive
   c. Expository
   d. Technical

37. Which of the following statements most accurately summarizes the passage?
   a. Tobacco is less healthy than many alternatives.
   b. Tobacco is deadly, expensive, and socially unappealing, and smokers would be much better off kicking the addiction.
   c. In the United States, local, state, and federal governments typically tax tobacco products, which leads to high prices.
   d. Tobacco products shorten smokers' lives by ten years and kill more than six million people per year.

38. The author would be most likely to agree with which of the following statements?
   a. Smokers should only quit cold turkey and avoid all nicotine cessation devices.
   b. Other substances are more addictive than tobacco.
   c. Smokers should quit for whatever reason that gets them to stop smoking.
   d. People who want to continue smoking should advocate for a reduction in tobacco product taxes.

39. Which of the following represents an opinion statement on the part of the author?
   a. According to the Centers for Disease Control and Prevention (CDC), tobacco products cause nearly six million deaths per year.
   b. Nicotine users who struggle with addiction sometimes pay more for a pack of cigarettes than a few gallons of gas.
   c. They also risk staining their teeth and fingers with yellow residue from the tar.
   d. Additionally, smokers tend to stink. The smell of smoke is all-consuming and creates a pervasive nastiness.

Questions 40 – 45 are based on the following passage:

What's About to Happen to Mr. Button?

As long ago as 1860 it was the proper thing to be born at home. At present, so I am told, the high gods of medicine have decreed that the first cries of the young shall be uttered upon the anesthetic air of a hospital, preferably a fashionable one. So young Mr. and Mrs. Roger Button were fifty years ahead of style when they decided, one day in the summer of 1860, that their first baby should be born in a hospital. Whether this anachronism had any bearing upon the astonishing history I am about to set down will never be known.

I shall tell you what occurred, and let you judge for yourself.

The Roger Buttons held an enviable position, both social and financial, in ante-bellum Baltimore. They were related to the This Family and the That Family, which, as every Southerner knew, entitled them to membership in that enormous peerage which largely populated the Confederacy. This was their first experience with the charming old custom of having babies—Mr. Button was naturally nervous. He hoped it would be a boy so that he could be sent to Yale College in Connecticut, at which institution Mr. Button himself had been known for four years by the somewhat obvious nickname of "Cuff."

On the September morning consecrated to the enormous event he arose nervously at six o'clock dressed himself, adjusted an impeccable stock, and hurried forth through the streets of Baltimore to the hospital, to determine whether the darkness of the night had borne in new life upon its bosom.

When he was approximately a hundred yards from the Maryland Private Hospital for Ladies and Gentlemen he saw Doctor Keene, the family physician, descending the front steps, rubbing his hands together with a washing movement—as all doctors are required to do by the unwritten ethics of their profession.

Mr. Roger Button, the president of Roger Button & Co., Wholesale Hardware, began to run toward Doctor Keene with much less dignity than was expected from a Southern gentleman of that picturesque period. "Doctor Keene!" he called. "Oh, Doctor Keene!"

The doctor heard him, faced around, and stood waiting, a curious expression settling on his harsh, medicinal face as Mr. Button drew near.

"What happened?" demanded Mr. Button, as he came up in a gasping rush. "What was it? How is she? A boy? Who is it? What—"

"Talk sense!" said Doctor Keene sharply. He appeared somewhat irritated.

"Is the child born?" begged Mr. Button.

Doctor Keene frowned. "Why, yes, I suppose so—after a fashion." Again he threw a curious glance at Mr. Button.

The Curious Case of Benjamin Button, F.S. Fitzgerald, 1922

40. What major event is about to happen in this story?
    a. Mr. Button is about to go to a funeral.
    b. Mr. Button's wife is about to have a baby.
    c. Mr. Button is getting ready to go to the doctor's office.
    d. Mr. Button is about to go shopping for new clothes.

41. What kind of tone does the above passage have?
    a. Nervous and Excited
    b. Sad and Angry
    c. Shameful and Confused
    d. Grateful and Joyous

42. What is the meaning of the word "consecrated" in paragraph 4?
    a. Numbed
    b. Chained
    c. Dedicated
    d. Moved

43. What does the author mean to do by adding the following statement?

"rubbing his hands together with a washing movement—as all doctors are required to do by the unwritten ethics of their profession."

a. Suggesting that Mr. Button is tired of the doctor.
b. Trying to explain the detail of the doctor's profession.
c. Hinting to readers that the doctor is an unethical man.
d. Giving readers a visual picture of what the doctor is doing.

44. Which of the following best describes the development of this passage?
a. It starts in the middle of a narrative in order to transition smoothly to a conclusion.
b. It is a chronological narrative from beginning to end.
c. The sequence of events is backwards—we go from future events to past events.
d. To introduce the setting of the story and its characters.

45. Which of the following is an example of an imperative sentence?
a. "Oh, Doctor Keene!"
b. "Talk sense!"
c. "Is the child born?"
d. "Why, yes, I suppose so—"

Questions 46 – 51 are based on the following passage:

Death or Freedom?

Knowing that Mrs. Mallard was afflicted with heart trouble, great care was taken to break to her as gently as possible the news of her husband's death.

It was her sister Josephine who told her, in broken sentences; veiled hints that revealed in half concealing. Her husband's friend Richards was there, too, near her. It was he who had been in the newspaper office when intelligence of the railroad disaster was received, with Brently Mallard's name leading the list of "killed." He had only taken the time to assure himself of its truth by a second telegram, and had hastened to forestall any less careful, less tender friend in bearing the sad message.

She did not hear the story as many women have heard the same, with a paralyzed inability to accept its significance. She wept at once, with sudden, wild abandonment, in her sister's arms. When the storm of grief had spent itself she went away to her room alone. She would have no one follow her.

There stood, facing the open window, a comfortable, roomy armchair. Into this she sank, pressed down by a physical exhaustion that haunted her body and seemed to reach into her soul.

She could see in the open square before her house the tops of trees that were all aquiver with the new spring life. The delicious breath of rain was in the air. In the street below a peddler was crying his wares. The notes of a distant song which some one was singing reached her faintly, and countless sparrows were twittering in the eaves.

There were patches of blue sky showing here and there through the clouds that had met and piled one above the other in the west facing her window.

She sat with her head thrown back upon the cushion of the chair, quite motionless, except when a sob came up into her throat and shook her, as a child who has cried itself to sleep continues to sob in its dreams.

She was young, with a fair, calm face, whose lines bespoke repression and even a certain strength. But now here was a dull stare in her eyes, whose gaze was fixed away off yonder on one of those patches of blue sky. It was not a glance of reflection, but rather indicated a suspension of intelligent thought.

There was something coming to her and she was waiting for it, fearfully. What was it? She did not know; it was too subtle and elusive to name. But she felt it, creeping out of the sky, reaching toward her through the sounds, the scents, and color that filled the air.

Now her bosom rose and fell tumultuously. She was beginning to recognize this thing that was approaching to possess her, and she was striving to beat it back with her will—as powerless as her two white slender hands would have been. When she abandoned herself a little whispered word escaped her slightly parted lips. She said it over and over under her breath: "free, free, free!" The vacant stare and the look of terror that had followed it went from her eyes. They stayed keen and bright. Her pulses beat fast, and the coursing blood warmed and relaxed every inch of her body.

She did not stop to ask if it were or were not a monstrous joy that held her. A clear and exalted perception enabled her to dismiss the suggestion as trivial. She knew that she would weep again when she saw the kind, tender hands folded in death; the face that had never looked save with love upon her, fixed and gray and dead. But she saw beyond that bitter moment a long procession of years to come that would belong to her absolutely. And she opened and spread her arms out to them in welcome.

Excerpt from "The Story of An Hour," Kate Chopin, 1894

46. What point of view is the above passage told in?
    a. First person
    b. Second person
    c. Third person omniscient
    d. Third person limited

47. What kind of irony are we presented with in this story?
    a. The way Mrs. Mallard reacted to her husband's death.
    b. The way in which Mr. Mallard died.
    c. The way in which the news of her husband's death was presented to Mrs. Mallard.
    d. The way in which nature is compared with death in the story.

48. What is the meaning of the word "elusive" in paragraph 9?
    a. Horrible
    b. Indefinable
    c. Quiet
    d. Joyful

49. What is the best summary of the passage above?
    a. Mr. Mallard, a soldier during World War I, is killed by the enemy and leaves his wife widowed.
    b. Mrs. Mallard understands the value of friendship when her friends show up for her after her husband's death.
    c. Mrs. Mallard combats mental illness daily and will perhaps be sent to a mental institution soon.
    d. Mrs. Mallard, a newly widowed woman, finds unexpected relief in her husband's death.

50. What is the tone of this story?
    a. Confused
    b. Joyful
    c. Depressive
    d. All of the above

51. What is the meaning of the word "tumultuously" in paragraph 10?
    a. Orderly
    b. Unashamedly
    c. Violently
    d. Calmly

Questions 52 – 56 are based upon the following passage:

This excerpt is an adaptation from Abraham Lincoln's Address Delivered at the Dedication of the Cemetery at Gettysburg, November 19, 1863.

How Can We Honor Them?

Four score and seven years ago our fathers brought forth on this continent, a new nation, conceived in liberty, and dedicated to the proposition that all men are created equal.

Now we are engaged in a great civil war, testing whether that nation, or any nation so conceived and so dedicated, can long endure. We are met on a great battlefield of that war. We have come to dedicate a portion of that field, as a final resting place for those who here gave their lives that this nation might live. It is altogether fitting and proper that we should do this.

But, in a larger sense, we cannot dedicate—we cannot consecrate that we cannot hallow—this ground. The brave men, living and dead, who struggled here, have consecrated it, far above our poor power to add or detract. The world will little note, nor long remember what we say here, but it can never forget what they did here. It is for us the living, rather, to be dedicated here to the unfinished work which they who fought here have thus far so nobly advanced. It is rather for us to be here and dedicated to the great task remaining before us—that from these honored dead we take increased devotion to that cause for which they gave the last full measure of devotion—that we here highly resolve that these dead shall not have died in vain—that this nation, under God, shall have a new birth of freedom—and that government of people, by the people, for the people, shall not perish from the earth.

52. The best description for the phrase "Four score and seven years ago" is?
    a. A unit of measurement
    b. A period of time
    c. A literary movement
    d. A statement of political reform

53. What is the setting of this text?
   a. A battleship off of the coast of France
   b. A desert plain on the Sahara Desert
   c. A battlefield in a North American town
   d. The residence of Abraham Lincoln

54. Which war is Abraham Lincoln referring to in the following passage?

> Now we are engaged in a great civil war, testing whether that nation, or any nation so conceived and so dedicated, can long endure.

   a. World War I
   b. The War of Spanish Succession
   c. World War II
   d. The American Civil War

55. What message is the author trying to convey through this address?
   a. The audience should perpetuate the ideals of freedom that the soldiers died fighting for.
   b. The audience should honor the dead by establishing an annual memorial service.
   c. The audience should form a militia that would overturn the current political structure.
   d. The audience should forget the lives that were lost and discredit the soldiers.

56. What is the effect of Lincoln's statement in the following passage?

> But, in a larger sense, we cannot dedicate—we cannot consecrate that we cannot hallow—this ground. The brave men, living and dead, who struggled here, have consecrated it, far above our poor power to add or detract.

   a. His comparison emphasizes the great sacrifice of the soldiers who fought in the war.
   b. His comparison serves as a reminder of the inadequacies of his audience.
   c. His comparison serves as a catalyst for guilt and shame among audience members.
   d. His comparison attempts to illuminate the great differences between soldiers and civilians.

Questions 57 – 59 are based on the following passage:

Who was George Washington?

George Washington emerged out of the American Revolution as an unlikely champion of liberty. On June 14, 1775, the Second Continental Congress created the Continental Army, and John Adams, serving in the Congress, nominated Washington to be its first commander. Washington fought under the British during the French and Indian War, and his experience and prestige proved instrumental to the American war effort. Washington provided invaluable leadership, training, and strategy during the Revolutionary War. He emerged from the war as the embodiment of liberty and freedom from tyranny.

After vanquishing the heavily favored British forces, Washington could have pronounced himself as the autocratic leader of the former colonies without any opposition, but he famously refused and returned to his Mount Vernon plantation. His restraint proved his commitment to the fledgling state's republicanism. Washington was later unanimously elected as the first American president. But it is Washington's farewell address that cemented his legacy as a visionary worthy of study.

In 1796, President Washington issued his farewell address by public letter. Washington enlisted his good friend, Alexander Hamilton, in drafting his most famous address. The letter expressed Washington's faith in the Constitution and rule of law. He encouraged his fellow Americans to put aside partisan differences and establish a national union. Washington warned Americans against meddling in foreign affairs and entering military alliances. Additionally, he stated his opposition to national political parties, which he considered partisan and counterproductive.

Americans would be wise to remember Washington's farewell, especially during presidential elections when politics hits a fever pitch. They might want to question the political institutions that were not planned by the Founding Fathers, such as the nomination process and political parties themselves.

57. Which of the following statements is logically based on the information contained in the passage above?
    a. George Washington's background as a wealthy landholder directly led to his faith in equality, liberty, and democracy.
    b. George Washington would have opposed America's involvement in the Second World War.
    c. George Washington would not have been able to write as great a farewell address without the assistance of Alexander Hamilton.
    d. George Washington would probably not approve of modern political parties.

58. Which of the following statements is the best description of the author's purpose in writing this passage about George Washington?
    a. To inform American voters about a Founding Father's sage advice on a contemporary issue and explain its applicability to modern times
    b. To introduce George Washington to readers as a historical figure worthy of study
    c. To note that George Washington was more than a famous military hero
    d. To convince readers that George Washington is a hero of republicanism and liberty

59. In which of the following materials would the author be the most likely to include this passage?
a. A history textbook
b. An obituary
c. A fictional story
d. A newspaper editorial

Questions 60 – 64 are based on the following passage:

Who First Came to the New World?

Christopher Columbus is often credited for discovering America. This is incorrect. First, it is impossible to "discover" something where people already live; however, Christopher Columbus did explore places in the New World that were previously untouched by Europe, so the term "explorer" would be more accurate. Another correction must be made, as well: Christopher Columbus was not the first European explorer to reach the present day Americas! Rather, it was Leif Erikson who first came to the New World and contacted the natives, nearly five hundred years before Christopher Columbus.

Leif Erikson, the son of Erik the Red (a famous Viking outlaw and explorer in his own right), was born in either 970 or 980, depending on which historian you seek. His own family, though, did not raise Leif, which was a Viking tradition. Instead, one of Erik's prisoners taught Leif reading and writing, languages, sailing, and weaponry. At age 12, Leif was considered a man and returned to his family. He killed a man during a dispute shortly after his return, and the council banished the Erikson clan to Greenland.

In 999, Leif left Greenland and traveled to Norway where he would serve as a guard to King Olaf Tryggvason. It was there that he became a convert to Christianity. Leif later tried to return home with the intention of taking supplies and spreading Christianity to Greenland, however his ship was blown off course and he arrived in a strange new land: present day Newfoundland, Canada.

When he finally returned to his adopted homeland Greenland, Leif consulted with a merchant who had also seen the shores of this previously unknown land we now know as Canada. The son of the legendary Viking explorer then gathered a crew of 35 men and set sail. Leif became the first European to touch foot in the New World as he explored present-day Baffin Island and Labrador, Canada. His crew called the land Vinland since it was plentiful with grapes.

During their time in present-day Newfoundland, Leif's expedition made contact with the natives whom they referred to as Skraelings (which translates to "wretched ones" in Norse). There are several secondhand accounts of their meetings. Some contemporaries described trade between the peoples. Other accounts describe clashes where the Skraelings defeated the Viking explorers with long spears, while still others claim the Vikings dominated the natives. Regardless of the circumstances, it seems that the Vikings made contact of some kind. This happened around 1000, nearly five hundred years before Columbus famously sailed the ocean blue.

Eventually, in 1003, Leif set sail for home and arrived at Greenland with a ship full of timber.

In 1020, seventeen years later, the legendary Viking died. Many believe that Leif Erikson should receive more credit for his contributions in exploring the New World.

60. Which of the following best describes how the author generally presents the information?
    a. Chronological order
    b. Comparison-contrast
    c. Cause-effect
    d. Conclusion-premises

61. Which of the following is an opinion, rather than historical fact, expressed by the author?
    a. Leif Erikson was definitely the son of Erik the Red; however, historians debate the year of his birth.
    b. Leif Erikson's crew called the land Vinland since it was plentiful with grapes.
    c. Leif Erikson deserves more credit for his contributions in exploring the New World.
    d. Leif Erikson explored the Americas nearly five hundred years before Christopher Columbus.

62. Which of the following most accurately describes the author's main conclusion?
    a. Leif Erikson is a legendary Viking explorer.
    b. Leif Erikson deserves more credit for exploring America hundreds of years before Columbus.
    c. Spreading Christianity motivated Leif Erikson's expeditions more than any other factor.
    d. Leif Erikson contacted the natives nearly five hundred years before Columbus.

63. Which of the following best describes the author's intent in the passage?
    a. To entertain
    b. To inform
    c. To alert
    d. To suggest

64. Which of the following can be logically inferred from the passage?
    a. The Vikings disliked exploring the New World.
    b. Leif Erikson's banishment from Iceland led to his exploration of present-day Canada.
    c. Leif Erikson never shared his stories of exploration with the King of Norway.
    d. Historians have difficulty definitively pinpointing events in the Vikings' history.

Questions 65 – 68 are based on the following passage:

Caribbean Island Destinations

Do you want to vacation at a Caribbean island destination? Who wouldn't want a tropical vacation? Visit one of the many Caribbean islands where visitors can swim in crystal blue waters, swim with dolphins, or enjoy family-friendly or adult-only resorts and activities. Every island offers a unique and picturesque vacation destination. Choose from these islands: Aruba, St. Lucia, Barbados, Anguilla, St. John, and so many more. A Caribbean island destination will be the best and most refreshing vacation ever . . . no regrets!

65. What is the topic of the passage?
    a. Caribbean island destinations
    b. Tropical vacation
    c. Resorts
    d. Activities

66. What is/are the supporting detail(s) of this passage?
    a. Cruising to the Caribbean
    b. Local events
    c. Family or adult-only resorts and activities
    d. All of the above

67. Read the following sentence, and answer the question below.
    "A Caribbean island destination will be the best and most refreshing vacation ever … no regrets!"

What is this sentence an example of?
    a. Fact
    b. Opinion
    c. Device
    d. Fallacy

68. What is the author's purpose of this passage?
    a. Entertain readers
    b. Persuade readers
    c. Inform readers
    d. None of the above

Questions 68 & 70 are based on the following passage:

Lola: The Siberian Husky

Meet Lola . . . Lola is an overly friendly Siberian husky who loves her long walks, digs holes for days, and sheds unbelievably . . . like a typical Siberian husky. Lola has to be brushed and brushed and brushed—did I mention that she has to be brushed . . . all the time! On her long walks, Lola loves making friends with new dogs and kids. A robber could break into our house, and even though they may be intimidated by Lola's wolf-like appearance, the robber would be shocked to learn that Lola would most likely greet them with kisses and a tail wag . . . she makes friends with everyone! Out of all the dogs we've ever owned, Lola is certainly one of a kind in many ways.

69. Based on the passage, what does the author imply?
    a. Siberian huskies are great pets but require a lot of time and energy.
    b. Siberian huskies are easy to take care of.
    c. Siberian huskies should not be around children.
    d. Siberian huskies are good guard dogs.

70. Because of their own experience with Siberian huskies, the author of the passage may be described as which of the following?
    a. Impartial
    b. Hasty
    c. Biased
    d. Irrational

Question 71 is based on the following passage:

While scientists aren't entirely certain why tornadoes form, they have some clues into the process. Tornadoes are dangerous funnel clouds that occur during a large thunderstorm. When warm, humid air near the ground meets cold, dry air from above, a column of the warm air can be drawn up into the clouds. Winds at different altitudes blowing at different speeds make the column of air rotate. As the spinning column of air picks up speed, a funnel cloud is formed. This funnel cloud moves rapidly and haphazardly. Rain and hail inside the cloud cause it to touch down, creating a tornado. Tornadoes move in a rapid and unpredictable pattern, making them extremely destructive and dangerous. Scientists continue to study tornadoes to improve radar detection and warning times.

71. The main purpose of this passage is to:
    a. Show why tornadoes are dangerous
    b. Explain how a tornado forms
    c. Compare thunderstorms to tornadoes
    d. Explain what to do in the event of a tornado

Question 72 is based on the following passage:

There are two major kinds of cameras on the market right now for amateur photographers. Camera enthusiasts can either purchase a digital single-lens reflex camera (DSLR) camera or a compact system camera (CSC). The main difference between a DSLR and a CSC is that the DSLR has a full-sized sensor, which means it fits in a much larger body. The CSC uses a mirrorless system, which makes for a lighter, smaller camera. While both take quality pictures, the DSLR generally has better picture quality due to the larger sensor. CSCs still take very good quality pictures and are more convenient to carry than a

DSLR. This makes the CSC an ideal choice for the amateur photographer looking to step up from a point-and-shoot camera.

72. The main difference between the DSLR and CSC is:
    a. The picture quality is better in the DSLR.
    b. The CSC is less expensive than the DSLR.
    c. The DSLR is a better choice for amateur photographers.
    d. The DSLR's larger sensor makes it a bigger camera than the CSC.

Question 73 is based on the following passage:

When selecting a career path, it's important to explore the various options available. Many students entering college may shy away from a major because they don't know much about it. For example, many students won't opt for a career as an actuary, because they aren't exactly sure what it entails. They would be missing out on a career that is very lucrative and in high demand. Actuaries work in the insurance field and assess risks and premiums. The average salary of an actuary is $100,000 per year. Another career option students may avoid, due to lack of knowledge of the field, is a hospitalist. This is a physician that specializes in the care of patients in a hospital, as opposed to those seen in private practices. The average salary of a hospitalist is upwards of $200,000. It pays to do some digging and find out more about these lesser-known career fields.

73. An actuary is:
    a. A doctor who works in a hospital
    b. The same as a hospitalist
    c. An insurance agent who works in a hospital
    d. A person who assesses insurance risks and premiums

Question 74 is based on the following passage:

Many people are unsure of exactly how the digestive system works. Digestion begins in the mouth where teeth grind up food and saliva breaks it down, making it easier for the body to absorb. Next, the food moves to the esophagus, and it is pushed into the stomach. The stomach is where food is stored and broken down further by acids and digestive enzymes, preparing it for passage into the intestines. The small intestine is where the nutrients are taken from food and passed into the blood stream. Other essential organs like the liver, gall bladder, and pancreas aid the stomach in breaking down food and absorbing nutrients. Finally, food waste is passed into the large intestine where it is eliminated by the body.

74. The purpose of this passage is to:
    a. Explain how the liver works.
    b. Show why it is important to eat healthy foods
    c. Explain how the digestive system works
    d. Show how nutrients are absorbed by the small intestine

Question 75 is based on the following passage:

Hard water occurs when rainwater mixes with minerals from rock and soil. Hard water has a high mineral count, including calcium and magnesium. The mineral deposits from hard water can stain hard surfaces in bathrooms and kitchens as well as clog pipes. Hard water can stain dishes, ruin clothes, and reduce the life of any appliances it touches, such as hot water heaters, washing machines, and humidifiers.

One solution is to install a water softener to reduce the mineral content of water, but this can be costly. Running vinegar through pipes and appliances and using vinegar to clean hard surfaces can also help with mineral deposits.

75. From this passage, it can be concluded that:
   a. Hard water can cause a lot of problems for homeowners.
   b. Calcium is good for pipes and hard surfaces.
   c. Water softeners are easy to install.
   d. Vinegar is the only solution to hard water problems.

Question 76 is based on the following passage:

Osteoporosis is a medical condition that occurs when the body loses bone or makes too little bone. This can lead to brittle, fragile bones that easily break. Bones are already porous, and when osteoporosis sets in, the spaces in bones become much larger, causing them to weaken. Both men and women can contract osteoporosis, though it is most common in women over age 50. Loss of bone can be silent and progressive, so it is important to be proactive in prevention of the disease.

76. The main purpose of this passage is to:
   a. Discuss some of the ways people contract osteoporosis
   b. Describe different treatment options for those with osteoporosis
   c. Explain how to prevent osteoporosis
   d. Define osteoporosis

Question 77 is based on the following passage:

Vacationers looking for a perfect experience should opt out of Disney parks and try a trip on Disney Cruise Lines. While a park offers rides, characters, and show experiences, it also includes long lines, often very hot weather, and enormous crowds. A Disney Cruise, on the other hand, is a relaxing, luxurious vacation that includes many of the same experiences as the parks, minus the crowds and lines. The cruise has top-notch food, maid service, water slides, multiple pools, Broadway-quality shows, and daily character experiences for kids. There are also many activities, such as bingo, trivia contests, and dance parties that can entertain guests of all ages. The cruise even stops at Disney's private island for a beach barbecue with characters, waterslides, and water sports. Those looking for the Disney experience without the hassle should book a Disney cruise.

77. The main purpose of this passage is to:
   a. Explain how to book a Disney cruise
   b. Show what Disney parks have to offer
   c. Show why Disney parks are expensive
   d. Compare Disney parks to the Disney cruise

Question 78 is based on the following passage:

Coaches of kids' sports teams are increasingly concerned about the behavior of parents at games. Parents are screaming and cursing at coaches, officials, players, and other parents. Physical fights have even broken out at games. Parents need to be reminded that coaches are volunteers, giving up their time and energy to help kids develop in their chosen sport. The goal of kids' sports teams is to learn and develop skills, but it's also to have fun. When parents are out of control at games and practices, it takes the fun out of the sport.

78. From this passage, it can be concluded that:
   a. Coaches are modeling good behavior for kids.
   b. Organized sports are not good for kids.
   c. Parents' behavior at their kids' games needs to change.
   d. Parents and coaches need to work together.

Question 79 is based on the following passage:

As summer approaches, drowning incidents will increase. Drowning happens very quickly and silently. Most people assume that drowning is easy to spot, but a person who is drowning doesn't make noise or wave his arms. Instead, he will have his head back and his mouth open, with just his face out of the water. A person who is truly in danger of drowning is not able to wave his arms in the air or move much at all. Recognizing these signs of drowning can prevent tragedy.

79. The main purpose of this passage is to:
   a. Explain the dangers of swimming
   b. Show how to identify the signs of drowning
   c. Explain how to be a lifeguard
   d. Compare the signs of drowning

Question 80 is based on the following passage:

Technology has been invading cars for the last several years, but there are some new high tech trends that are pretty amazing. It is now standard in many car models to have a rear-view camera, hands-free phone and text, and a touch screen digital display. Music can be streamed from a paired cell phone, and some displays can even be programmed with a personal photo. Sensors beep to indicate there is something in the driver's path when reversing and changing lanes. Rain-sensing windshield wipers and lights are automatic, leaving the driver with little to do but watch the road and enjoy the ride. The next wave of technology will include cars that automatically parallel park, and a self-driving car is on the horizon. These technological advances make it a good time to be a driver.

80. It can be concluded from this paragraph that:
   a. Technology will continue to influence how cars are made.
   b. Windshield wipers and lights are always automatic.
   c. It is standard to have a rear-view camera in all cars.
   d. Technology has reached its peak in cars.

Questions 81 – 83 are based upon the following passage:

This excerpt is adapted from "The 'Hatchery' of the Sun-Fish"--- Scientific American, #711

I have thought that an example of the intelligence (instinct?) of a class of fish which has come under my observation during my excursions into the Adirondack region of New York State might possibly be of interest to your readers, especially as I am not aware that any one except myself has noticed it, or, at least, has given it publicity.

The female sun-fish (called, I believe, in England, the roach or bream) makes a "hatchery" for her eggs in this wise. Selecting a spot near the banks of the numerous lakes in which this region abounds, and where the water is about 4 inches deep, and still, she builds, with her tail and snout, a circular embankment 3 inches in height and 2 thick. The circle, which is as perfect a one as could be formed with mathematical instruments, is usually a foot and a half in diameter; and at one side of this circular wall an opening is left by the fish of just sufficient width to admit her body.

The mother sun-fish, having now built or provided her "hatchery," deposits her spawn within the circular inclosure, and mounts guard at the entrance until the fry are hatched out and are sufficiently large to take charge of themselves. As the embankment, moreover, is built up to the surface of the water, no enemy can very easily obtain an entrance within the inclosure from the top; while there being only one entrance, the fish is able, with comparative ease, to keep out all intruders.

I have, as I say, noticed this beautiful instinct of the sun-fish for the perpetuity of her species more particularly in the lakes of this region; but doubtless the same habit is common to these fish in other waters.

81. What is the purpose of this passage?
    a. To show the effects of fish hatcheries on the Adirondack region
    b. To persuade the audience to study Ichthyology (fish science)
    c. To depict the sequence of mating among sun-fish
    d. To enlighten the audience on the habits of sun-fish and their hatcheries

82. What does the word *wise* in this passage most closely mean?
    a. Knowledge
    b. Manner
    c. Shrewd
    d. Ignorance

83. What is the definition of the word *fry* as it appears in the following passage?

    The mother sun-fish, having now built or provided her "hatchery," deposits her spawn within the circular inclosure, and mounts guard at the entrance until the fry are hatched out and are sufficiently large to take charge of themselves.

    a. Fish at the stage of development where they are capable of feeding themselves.
    b. Fish eggs that have been fertilized.
    c. A place where larvae is kept out of danger from other predators.
    d. A dish where fish is placed in oil and fried until golden brown.

Question 84 is based on the following passage.

A famous children's author recently published a historical fiction novel under a pseudonym; however, it did not sell as many copies as her children's books. In her earlier years, she had majored in history and earned a graduate degree in Antebellum American History, which is the time frame of her new novel. Critics praised this newest work far more than the children's series that made her famous. In fact, her new novel was nominated for the prestigious Albert J. Beveridge Award but still isn't selling like her children's books, which fly off the shelves because of her name alone.

84. Which one of the following statements might be accurately inferred based on the above passage?
   a. The famous children’s author produced an inferior book under her pseudonym.
   b. The famous children’s author is the foremost expert on Antebellum America.
   c. The famous children’s author did not receive the bump in publicity for her historical novel that it would have received if it were written under her given name.
   d. People generally prefer to read children's series than historical fiction.

Answer Explanations

1. A: Instigate refers to initiating or bringing about an action. While instigating something can be overwhelming, overwhelm describes the action, not what the action is. None of these choices match this except for begin, which is synonymous with instigate.

2. C: From the Latin root *fortis*, fortify literally means to make strong. Foster has nothing to do with the term at all. Undermine is to weaken, so this is the opposite. While a barricade can be used to fortify something, it is a means of protection, not the action of protection, unless used as a verb.

3. D: An ultimatum is a final demand or statement that carries a consequence if not met. The closest term presented is threat because, in context, an ultimatum carries some kind of punitive action or penalty if the terms are not met. Generally, it is usually used in a threatening or forceful context.

4. A: This was a tricky question because some choices connect wave to actions involved with water. Sink and tide are actions associated with the water, but are distinct from an ocean wave, which means wave is not used in context of an ocean wave, but as a verb to wave. To wave is to move around. Stagnant is to be still. Flourish describes a waving motion, making it the correct match.

5. C: It's easy to confuse void with avoid, which means keep away from. Void describes nothingness or emptiness. This means that around *(A)* is not the correct answer. Therefore, the best choice is empty, because this is the meaning of void.

6. C: Query comes from the Latin *querere*, which means to ask, so query means to ask a question. This Latin root also appears in the word question, which gives you a clue that the act of seeking answers is involved. Ask has the same meaning as query.

7. B: Severe reflects high intensity or very great. Cut and shake are verbs with actions that do not reflect severe. While something can be severely foul, foul is a broad description of something bad, but not necessarily of a high level of intensity. Extreme *(B)* describes something of the highest or most serious nature.

8. D: Vexation describes a state of irritation or annoyance. Think of the term vexed, meaning annoyed. The closest choice is anger, because vexation reflects annoyance, which, in most contexts, means a person is slightly angered or feels mild anger.

9. A: Loquacious reflects the tendency to talk a lot. While someone can be humorous and loquacious, humorous describes the kind of talk, not the fact that someone talks a lot. Cranky has nothing to do with the word, and silent is a clear opposite. This leaves verbose, meaning using an abundance of words; loquacious is a synonym.

10. B: Grim describes something of a dark or foreboding nature. Frank reflects sincerity, so does not relate. Neither does corpse, which is a body. Sharp describes a witty response or edged surface. This leaves dire, which is synonymous with grim. Both words reflect dark or unfruitful circumstances.

11. B: To disparage someone is to put them down. Although this may involve breaking a spirit and causing despair, despair is the result of disparage. To spar or fight may also be a result of disparage. Belittle means to bring someone down with words and is synonymous with disparage.

12. D: The other choices refer to attitude, a word spelled and pronounced similarly to altitude. Someone's attitude can reflect haughtiness, influence their outlook, and reflect their behavior. None of these terms describe altitude, which is the measure of height. Height is directly related to altitude.

13. A: Looking at the root word *ego*, egotistic must have something to do with the self – in this case, excessive self-interest. Such a person tends to be the opposite of altruistic, which means selfless. Unconcerned is also inappropriate, as egotistic people are concerned for themselves. Agrarian is an unrelated word concerning fields or farm lifestyle. Conceited is synonymous to egotistic.

14. B: Build and guilt can be ruled out because they are not related to understand. Endow is more difficult, because one can endow, or give, someone knowledge to understand, but endow involves the act of giving. Fathom is synonymous with understand; both terms reflect being able to grasp information.

15. D: Terrain, from the Latin *terra* refers to the earth or physical landscape. Celestial and firmament both describe the sky. Arboreal describes things relating to trees. This leaves ground, another word for Earth or land, the same as terrain.

16. A: Mitigate refers to easing tension or making less severe. Focus, fiery, and conspire do not relate. Alleviate is synonymous, meaning to make less severe. Note the ending *-ate*, which also indicates function. Both terms reflect the function of easing difficult circumstances.

17. C: Encompass means to hold within or surround. Avoid and retreat are opposites. Brazen has no relation. Surround, meaning to encircle, is synonymous with encompass.

18. D: An archetype is an example of a person or thing, a recurrent symbol. An ancestor may be a good model of ideal behavior, but the term refers to someone who is related. Archetypes appear in literature, but these are different terms. Policy is unrelated. A model is a standard or representative, which is synonymous with archetype.

19. B: Contrast means to go against or to have a different perspective. Revive and improve are unrelated. Partisan can mean favoring one side, but variation is the best choice. Variation indicates clear difference, something that is not uniform and, therefore, contrasting.

20. C: Irate comes from the Latin *ira-*, which gives it the meaning of angry or irritable. Amiable means happy and friendly. Innocent and tangible have different and unrelated meanings. Incensed comes from the Latin *incedere*, which means to burn. Often this burning is a metaphor for extreme anger, the meaning of irate. Thus, incensed and irate are synonyms.

21. A: Plausible means likely to be possible or accepted, the opposite of skeptical. Diurnal relates to daytime, so is unrelated. Erudite means clever or intelligent, but not necessarily possible or correct. Conceivable means that something is able to be thought of or able to be done. Conceivable is the best match for plausible.

22. C: Discourse refers to spoken and written communication, or debate. Path doesn't have anything to do with discourse, unless used figuratively. While discourse may consist of praise or disagreement, all discourse – written or verbal – is a form of speech. Therefore, speech is the best term that encompasses the same meaning as discourse.

23. A: The Latin root, *asper,* means rough. Exasperate, then, means to make relations with someone rough or to rub them the wrong way. Disappointment can be the result of an exasperating situation, but these are results of the term, not the same as exasperation itself. Annoy is a synonym, meaning to irritate and make someone slightly angry.

24. C: The Latin root *bell*, which comes from *bellum*, refers to war. Someone who is pugnacious is ready for a fight. One may be angry, patriotic, or loud, but none of these terms directly relate to warlike behavior like pugnacious.

25. D: Someone who displays equanimity, like the *equ-* prefix suggests, is level-headed and even-tempered. While equine and equanimity appear to share the *equ-* prefix, equine refers to horses. Justice and hostility don't relate at all. This leaves composure, which also describes one's ability to keep a calm and level-headed state.

26. A: Preemptive refers to an action taken before an anticipated result can occur, often as a preventive measure. Prepare is similar – it's defined as actions taken before an event – but it doesn't necessarily involve preventative measures as does preemptive. Preemptive measures can be hidden, but that describes the act. Initial – meaning first – is close because it can be the first action in a series, but again, it doesn't refer to an action that is preventative. Beforehand is an action done in advance, before something occurs.

27. D: Agitate means to annoy, which is not the same as the meaning in this sentence. Instigate can mean to start, which is not synonymous with liberate. Fracture is to crack or break something, which can metaphorically be attributed to liberation (breaking of chains), but is not directly related to the word liberate. From the Latin root *liber*, meaning free, liberate means to free or release. Release is synonymous.

28. B: Mediocre is from the Latin *medius*, meaning middle. It refers to something of only decent quality, not exceptional. This rules out excellent and inspiring, as both communicate ideas of surpassing quality. Intrusive is unrelated. Average means usual or nothing out of the ordinary, like mediocre.

29. C: Moot means uncertain or in dispute. We can eliminate factual, historical, and exemplify because these are common terms with no ties to moot. Debatable is defined as having uncertain circumstances, leading to discussion or dispute.

30. B: Bestow means to give or present. Take can be eliminated because to take is the opposite action of to give. Study and energize mean different things entirely. Bequeath means to give or leave to another person.

31. B: Barren means deserted, void, lifeless, or having little. A desert is barren because it produces little vegetation. Fruitful, sumptuous and lavish express richness and abundance, which contradict barren. Infertile means unable to produce life, which mirrors barren.

32. A: Refurbish is to restore, set things up again, or make repairs. Enlighten, craven, and burnish are unrelated to these ideas. Note instead the *re-* prefix in renovate, which means again. This gives refurbish and renovate a meaning of restoring again, or returning to a better state. In other words, renewal.

33. D: Chastise means to reprimand severely. Honor, choose, and locate are unrelated. Rebuke is defined as harshly disapproving someone. Both rebuke and chastise are verbs, making rebuke a match.

34. B: Sustain is to revitalize, or to give strength. Strength would be a good choice, but strength alone does not describe the re-strengthening that sustain embodies. The best choice is invigorate, which is a verb like sustain, meaning to strengthen. This makes invigorate the better choice. Expedite is unrelated. While embolden means to give someone courage, which is a form of strength, invigorate and sustain speak more toward physical circumstances.

35. A: Acquit means to free of blame or charge. While acquire and acquit appear similar, they are unrelated. Stall and quit are also unrelated. Forgive literally means to pardon of sins or offenses. While not appearing related by their spellings, their meanings are nearly a perfect match, and they are both verbs.

36. B: Narrative, Choice *A*, means a written account of connected events. Think of narrative writing as a story. Choice C, expository writing, generally seeks to explain or describe some phenomena, whereas Choice *D*, technical writing, includes directions, instructions, and/or explanations. This passage is persuasive writing, which hopes to change someone's beliefs based on an appeal to reason or emotion. The author is aiming to convince the reader that smoking is terrible. They use health, price, and beauty in their argument against smoking, so Choice *B*, persuasive, is the correct answer.

37. B: The author is opposed to tobacco. He cites disease and deaths associated with smoking. He points to the monetary expense and aesthetic costs. Choice *A* is wrong because alternatives to smoking are not even addressed in the passage. Choice *C* is wrong because it does not summarize the passage but rather is just a premise. Choice *D* is wrong because, while these statistics are a premise in the argument, they do not represent a summary of the piece. Choice *B* is the correct answer because it states the three critiques offered against tobacco and expresses the author's conclusion.

38. C: We are looking for something the author would agree with, so it should be anti-smoking or an argument in favor of quitting smoking. Choice *A* is wrong because the author does not speak against means of cessation. Choice *B* is wrong because the author does not reference other substances but does speak of how addictive nicotine, a drug in tobacco, is. Choice *D* is wrong because the author would not encourage reducing taxes to encourage a reduction of smoking costs, thereby helping smokers to continue the habit. Choice *C* is correct because the author is attempting to persuade smokers to quit smoking.

39. D: Here, we are looking for an opinion of the author's rather than a fact or statistic. Choice *A* is wrong because quoting statistics from the Centers of Disease Control and Prevention is stating facts, not opinions. Choice *B* is wrong because it expresses the fact that cigarettes sometimes cost more than a few gallons of gas. It would be an opinion if the author said that cigarettes were not affordable. Choice *C* is incorrect because yellow stains are a known possible adverse effect of smoking. Choice *D* is correct as an opinion because smell is subjective. Some people might like the smell of smoke, they might not have working olfactory senses, and/or some people might not find the smell of smoke akin to "pervasive nastiness," so this is the expression of an opinion. Thus, Choice *D* is the correct answer.

40. B: Mr. Button's wife is about to have a baby. The passage begins by giving the reader information about traditional birthing situations. Then, we are told that Mr. and Mrs. Button decide to go against tradition to have their baby in a hospital. The next few passages are dedicated to letting the reader know how Mr. Button dresses and goes to the hospital to welcome his new baby. There is a doctor in this excerpt, as Choice *C* indicates, and Mr. Button does put on clothes, as Choice *D* indicates. However, Mr. Button is not going to the doctor's office nor is he about to go shopping for new clothes.

41. A: The tone of the above passage is nervous and excited. We are told in the fourth paragraph that Mr. Button "arose nervously." We also see him running without caution to the doctor to find out about his wife and baby—this indicates his excitement. We also see him stuttering in a nervous yet excited fashion as he asks the doctor if it's a boy or girl. Though the doctor may seem a bit abrupt at the end, indicating a bit of anger or shame, neither of these choices is the overwhelming tone of the entire passage. Despite the circumstances, joy and gratitude are not the main tone in the passage.

42. C: Dedicated. Mr. Button is dedicated to the task before him. Choice *A*, numbed, Choice *B*, chained, and Choice *D*, moved, all could grammatically fit in the sentence. However, they are not synonyms with *consecrated* like Choice *C* is.

43. D: Giving readers a visual picture of what the doctor is doing. The author describes a visual image—the doctor rubbing his hands together—first and foremost. The author may be trying to make a comment about the profession; however, the author does not "explain the detail of the doctor's profession" as Choice *B* suggests.

44. D: To introduce the setting of the story and its characters. We know we are being introduced to the setting because we are given the year in the very first paragraph along with the season: "one day in the summer of 1860." This is a classic structure of an introduction of the setting. We are also getting a long explanation of Mr. Button, what his work is, who is related to him, and what his life is like in the third paragraph.

45. B: "Talk sense!" is an example of an imperative sentence. An imperative sentence gives a command. The doctor is commanding Mr. Button to talk sense. Choice *A* is an example of an exclamatory sentence, which expresses excitement. Choice *C* is an example of an interrogative sentence—these types of sentences ask questions. Choice *D* is an example of a declarative sentence. This means that the character is simply making a statement.

46. C: The point of view is told in third-person omniscient. We know this because the story starts out with us knowing something that the character does not know: that her husband has died. Mrs. Mallard eventually comes to know this, but we as readers know this information before it is broken to her. In third person limited, Choice *D*, we would only see and know what Mrs. Mallard herself knew, and we would find out the news of her husband's death when she found out the news, not before.

47. A: The way Mrs. Mallard reacted to her husband's death. The irony in this story is called situational irony, which means the situation that takes place is different than what the audience anticipated. At the beginning of the story, we see Mrs. Mallard react with a burst of grief to her husband's death. However, once she's alone, she begins to contemplate her future and says the word "free" over and over. This is quite a different reaction from Mrs. Mallard than what readers expected from the first of the story.

48. B: The word "elusive" most closely means "indefinable." Horrible, Choice *A*, doesn't quite fit with the tone of the word "subtle" that comes before it. Choice *C*, "quiet," is more closely related to the word "subtle." Choice *D*, "joyful," also doesn't quite fit the context here. "Indefinable" is the best option.

49. D: Mrs. Mallard, a newly widowed woman, finds unexpected relief in her husband's death. A summary is a brief explanation of the main point of a story. The story mostly focuses on Mrs. Mallard and her reaction to her husband's death, especially in the room when she's alone and contemplating the present and future. All of the other answer choices except Choice *C* are briefly mentioned in the story; however, they are not the main focus of the story.

50. D: The interesting thing about this story is that feelings that are confused, joyful, and depressive all play a unique and almost equal part of this story. There is no one right answer here, because the author seems to display all of these emotions through the character of Mrs. Mallard. She displays feelings of depressiveness by her grief at the beginning; then, when she receives feelings of joy, she feels moments of confusion. We as readers cannot help but go through these feelings with the character. Thus, the author creates a tone of depression, joy, and confusion, all in one story.

51. C: The word "tumultuously" most nearly means "violently." Even if you don't know the word "tumultuously," look at the surrounding context to figure it out. The next few sentences we see Mrs. Mallard striving to "beat back" the "thing that was approaching to possess her." We see a fearful and almost violent reaction to the emotion that she's having. Thus, her chest would rise and fall tumultuously, or violently.

52. B: A period of time. "Four score and seven years ago" is the equivalent of eighty-seven years, because the word "score" means "twenty." Choices *A* and *C* are incorrect because the context for describing a unit of measurement or a literary movement is lacking. *D* is incorrect because although Lincoln's speech is a cornerstone in political rhetoric, the phrase "Four score and seven years ago" is better narrowed to a period of time.

53. C: The setting of this text is a battlefield in Gettysburg, PA. Choices *A, B*, and *D* are incorrect because the text specifies that they "met on a great battlefield of that war."

54. D: Abraham Lincoln is the former president of the United States, so the correct answer is *D*, "The American Civil War." Though the U.S. was involved in World War I and II, Choices *A* and *C* are incorrect because a civil war specifically means citizens fighting within the same country. *B* is incorrect, as "The War of Spanish Succession" involved Spain, Italy, Germany, and Holland, and not the United States.

55. A: The speech calls on the audience to consider the soldiers who died on the battlefield as ideas to perpetuate freedom so that their deaths would not be in vain. Choice *B* is incorrect because, although they are there to "dedicate a portion of that field," there is no mention in the text of an annual memorial service. Choice *C* is incorrect because there is no charged language in the text, only reverence for the dead. Choice *D* is incorrect because "forget[ting] the lives that were lost" is the opposite of what Lincoln is suggesting.

56. A: Choice *A* is correct because Lincoln's intention was to memorialize the soldiers who had fallen as a result of war as well as celebrate those who had put their lives in danger for the sake of their country. Choices *B, C*, and *D* are incorrect because Lincoln's speech was supposed to foster a sense of pride among the members of the audience while connecting them to the soldiers' experiences, not to alienate or discourage them.

57. D: Although Washington was from a wealthy background, the passage does not say that his wealth led to his republican ideals, so Choice *A* is not supported. Choice *B* also does not follow from the passage. Washington's warning against meddling in foreign affairs does not mean that he would oppose wars of every kind, so Choice *B* is wrong. Choice *C* is also unjustified since the author does not indicate that Alexander Hamilton's assistance was absolutely necessary. Choice *D* is correct because the farewell address clearly opposes political parties and partisanship. The author then notes that presidential elections often hit a fever pitch of partisanship. Thus, it follows that George Washington would not approve of modern political parties and their involvement in presidential elections.

58. A: The author finishes the passage by applying Washington's farewell address to modern politics, so the purpose probably includes this application. Choice *B* is wrong because George Washington is already a well-established historical figure; furthermore, the passage does not seek to introduce him. Choice *C* is wrong because the author is not fighting a common perception that Washington was merely a military hero. Choice *D* is wrong because the author is not convincing readers. Persuasion does not correspond to the passage. Choice *A* states the primary purpose.

59. D: Choice *A* is wrong because the last paragraph is not appropriate for a history textbook. Choice *B* is false because the piece is not a notice or announcement of Washington's death. Choice *C* is clearly false because it is not fiction, but a historical writing. Choice *D* is correct. The passage is most likely to appear in a newspaper editorial because it cites information relevant and applicable to the present day, a popular format in editorials.

60. D: The passage does not proceed in chronological order since it begins by pointing out Christopher Columbus's explorations in America so Choice *A* does not work. Although the author compares and contrasts Erikson with Christopher Columbus, this is not the main way the information is presented; therefore, Choice *B* does not work. Neither does Choice *C* because there is no mention of or reference to cause and effect in the passage. However, the passage does offer a conclusion (Leif Erikson deserves more credit) and premises (first European to set foot in the New World and first to contact the natives) to substantiate Erikson's historical importance. Thus, Choice *D* is correct.

61. C: Choice *A* is wrong because it describes facts: Leif Erikson was the son of Erik the Red and historians debate Leif's date of birth. These are not opinions. Choice *B* is wrong; that Erikson called the land Vinland is a verifiable fact as is Choice *D* because he did contact the natives almost 500 years before Columbus. Choice *C* is the correct answer because it is the author's opinion that Erikson deserves more credit. That, in fact, is his conclusion in the piece, but another person could argue that Columbus or another explorer deserves more credit for opening up the New World to exploration. Rather than being an incontrovertible fact, it is a subjective value claim.

62. B: Choice *A* is wrong because the author aims to go beyond describing Erikson as a mere legendary Viking. Choice *C* is wrong because the author does not focus on Erikson's motivations, let alone name the spreading of Christianity as his primary objective. Choice *D* is wrong because it is a premise that Erikson contacted the natives 500 years before Columbus, which is simply a part of supporting the author's conclusion. Choice *B* is correct because, as stated in the previous answer, it accurately identifies the author's statement that Erikson deserves more credit than he has received for being the first European to explore the New World.

63. B: Choice *A* is wrong because the author is not in any way trying to entertain the reader. Choice *D* is wrong because he goes beyond a mere suggestion; "suggest" is too vague. Although the author is certainly trying to alert the readers of Leif Erikson's unheralded accomplishments, the nature of the writing does not indicate the author would be satisfied with the reader merely knowing of Erikson's exploration (Choice *C*). Rather, the author would want the reader to be informed about it, which is more substantial (Choice *B*).

64. D: Choice *A* is wrong because the author never addresses the Vikings' state of mind or emotions. Choice *B* is wrong because the author does not elaborate on Erikson's exile and whether he would have become an explorer if not for his banishment. Choice *C* is wrong because there is not enough information to support this premise. It is unclear whether Erikson informed the King of Norway of his finding. Although it is true that the King did not send a follow-up expedition, he could have simply chosen not to expend the resources after receiving Erikson's news. It is not possible to logically infer whether Erikson told him. Choice *D* is correct because there are two examples—Leif Erikson's date of birth and what happened during the encounter with the natives—of historians having trouble pinning down important details in Viking history.

65. A: The topic of the passage is Caribbean island destinations. The *topic* of the passage can be described in a one- or two-word phrase. Remember, when paraphrasing a passage, it is important to include the topic. Paraphrasing is when one puts a passage into his or her own words.

66. C: Family or adult-only resorts and activities are supporting details in this passage. *Supporting details* are details that help readers better understand the main idea. They answer questions such as who, what, where, when, why, or how. In this question, cruises and local events are not discussed in the passage, whereas family and adult-only resorts and activities support the main idea.

67. B: This sentence is an opinion. An *opinion* is when the author states his or her judgment or thoughts on a subject. In this example, the author implies that the reader will not regret the vacation and that it may be the best and most relaxing vacation, when in fact that may not be true. Therefore, the statement is the author's opinion. A fallacy is a flawed argument of mistaken belief based on faulty reasoning.

68. B: The author of the passage is trying to *persuade* readers to vacation in a Caribbean island destination by providing enticing evidence and a variety of options. The passage even includes the author's opinion. Not only does the author provide many details to support his or her opinion, the author also implies that the reader would almost be "in the wrong" if he or she didn't want to visit a Caribbean island, hence, the author is trying to persuade the reader to visit a Caribbean island.

69. A: The author implies that Siberian huskies are great pets but require a lot of time and energy. In the passage, the writer describes how huskies require lots of brushing and long walks and how they dig, making them not easy to care for. The author also describes how friendly Siberian huskies can be, even possibly greeting a robber at their own house, definitely not making them good guard dogs. Therefore, Siberian huskies are great pets but require a lot of time and energy.

70. C: The author may be *biased* because they show prejudice over one breed versus another in an unfair way. Impartial means fair, and is essentially the opposite of biased. Hasty means quick to judge and irrational means unreasonable or illogical.

71. B: The main point of this passage is to show how a tornado forms. Choice *A* is off base because while the passage does mention that tornadoes are dangerous, it is not the main focus of the passage. While thunderstorms are mentioned, they are not compared to tornadoes, so Choice *C* is incorrect. Choice *D* is incorrect because the passage does not discuss what to do in the event of a tornado.

72. D: The passage directly states that the larger sensor is the main difference between the two cameras. Choices *A* and *B* may be true, but these answers do not identify the major difference between the two cameras. Choice *C* states the opposite of what the paragraph suggests is the best option for amateur photographers, so it is incorrect.

73. D: An actuary assesses risks and sets insurance premiums. While an actuary does work in insurance, the passage does not suggest that actuaries have any affiliation with hospitalists or working in a hospital, so all other choices are incorrect.

74. C: The purpose of this passage is to explain how the digestive system works. Choice *A* focuses only on the liver, which is a small part of the process and not the focus of the paragraph. Choice *B* is off-track because the passage does not mention healthy foods. Choice *D* only focuses on one part of the digestive system.

75. A: The passage focuses mainly on the problems of hard water. Choice *B* is incorrect because calcium is not good for pipes and hard surfaces. The passage does not say anything about whether water softeners are easy to install, so *C* is incorrect. *D* is also incorrect because the passage does offer other solutions besides vinegar.

76. D: The main point of this passage is to define osteoporosis. Choice *A* is incorrect because the passage does not list ways that people contract osteoporosis. Choice *B* is incorrect because the passage does not mention any treatment options. While the passage does briefly mention prevention, it does not explain how, so Choice *C* is incorrect.

77. D: The passage compares Disney cruises with Disney parks. It does not discuss how to book a cruise, so Choice *A* is incorrect. Choice *B* is incorrect because though the passage does mention some of the park attractions, it is not the main point. The passage does not mention the cost of either option, so Choice *C* is incorrect.

78. C: The main point of this paragraph is that parents need to change their poor behavior at their kids' sporting events. Choice *A* is incorrect because the coaches' behavior is not mentioned in the paragraph. *B* suggests that sports are bad for kids, when the paragraph is about parents' behavior, so it is incorrect. While Choice *D* may be true, it offers a specific solution to the problem, which the paragraph does not discuss.

79. B: The point of this passage is to show what drowning looks like. Choice *A* is incorrect because while drowning is a danger of swimming, the passage doesn't include any other dangers. The passage is not intended for lifeguards specifically, but for a general audience, so Choice *C* is incorrect. There are a few signs of drowning, but the passage does not compare them; thus, Choice *D* is incorrect.

80. A: The passage discusses recent technological advances in cars and suggests that this trend will continue in the future with self-driving cars. Choice *B* and *C* are not true, so these are both incorrect. Choice *D* is also incorrect because the passage suggests continuing growth in technology, not a peak.

81. D: To enlighten the audience on the habits of sun-fish and their hatcheries. Choice *A* is incorrect because although the Adirondack region is mentioned in the text, there is no cause or effect relationships between the region and fish hatcheries depicted here. Choice *B* is incorrect because the text does not have an agenda, but rather is meant to inform the audience. Finally, Choice *C* is incorrect because the text says nothing of how sun-fish mate.

82. B: The word *wise* in this passage most closely means *manner*. Choices *A* and *C* are synonyms of *wise*; however, they are not relevant in the context of the text. Choice *D*, *ignorance*, is opposite of the word wise and is therefore incorrect.

83. A: Fish at the stage of development where they are capable of feeding themselves. Even if the word *fry* isn't immediately known to the reader, the context gives a hint when it says "until the fry are hatched out and are sufficiently large to take charge of themselves."

84. C: We are looking for an inference—a conclusion that is reached on the basis of evidence and reasoning—from the passage that will likely explain why the famous children's author did not achieve her usual success with the new genre (despite the book's acclaim). Choice *A* is wrong because the statement is false according to the passage. Choice *B* is wrong because, although the passage says the author has a graduate degree on the subject, it would be an unrealistic leap to infer that she is the foremost expert on Antebellum America. Choice *D* is wrong because there is nothing in the passage to lead us to infer that people generally prefer a children's series to historical fiction. In contrast, Choice *C* can be logically inferred since the passage speaks of the great success of the children's series and the declaration that the fame of the author's name causes the children's books to "fly off the shelves." Thus, she did not receive any bump from her name since she published the historical novel under a pseudonym, which makes Choice *C* correct.

Language

Mechanics

Usage

Spelling might or might not be important to some, or maybe it just doesn't come naturally, but those who are willing to discover some new ideas and consider their benefits can learn to spell better and improve their writing. Misspellings reduce a writer's credibility and can create misunderstandings. Spell checkers built into word processors are not a substitute for accuracy. They are neither foolproof nor without error. In addition, a writer's misspelling of one word may also be a word. For example, a writer intending to spell *herd* might accidentally type *s* instead of *d* and unintentionally spell *hers*. Since *her*s is a word, it would not be marked as a misspelling by a spell checker. In short, use spell check, but don't rely on it.

Guidelines for Spelling

Saying and listening to a word serves as the beginning of knowing how to spell it. Keep these subsequent guidelines in mind, remembering there are often exceptions because the English language is replete with them.

Guideline #1: Syllables must have at least one vowel. In fact, every syllable in every English word has a vowel.

- d*o*g
- h*a*yst*a*ck
- *a*nsw*e*r*i*ng
- *a*bst*e*nt*iou*s
- s*i*mpl*e*

Guideline #2: The long and short of it. When the vowel has a short vowel sound as in *mad* or *bed,* only the single vowel is needed. If the word has a long vowel sound, add another vowel, either alongside it or separated by a consonant: bed/*bead*; mad/*made.* When the second vowel is separated by two consonants— *madder*—it does not affect the first vowel's sound.

Guideline #3: Suffixes. Refer to the examples listed above.

Guideline #4: Which comes first; the *i* or the *e*? Remember the saying, "*I* before *e* except after *c* or when sounding as *a* as in *neighbor* or *weigh*." Keep in mind that these are only guidelines and that there are always exceptions to every rule.

Guideline #5: Vowels in the right order. Another helpful rhyme is, "When two vowels go walking, the first one does the talking." When two vowels are in a row, the first one often has a long vowel sound and the other is silent. An example is *team*.

If you have difficulty spelling words, determine a strategy to help. Work on spelling by playing word games like Scrabble or Words with Friends. Consider using phonics, which is sounding words out by slowly and surely stating each syllable. Try repeating and memorizing spellings as well as picturing words in your head. Try making up silly memory aids. See what works best.

Irregular Plurals

Irregular plurals are words that aren't made plural the usual way.

- Most nouns are made plural by adding *–s* (book*s*, television*s*, skyscraper*s*).
- Most nouns ending in *ch, sh, s, x,* or *z* are made plural by adding *–es* (church*es,* marsh*es*).
- Most nouns ending in a vowel + *y* are made plural by adding *–s* (day*s,* toy*s*).
- Most nouns ending in a consonant + *y,* are made plural by the *-y* becoming *-ies* (baby becomes *babies*).
- Most nouns ending in an *o* are made plural by adding *–s* (piano*s,* photo*s*).
- Some nouns ending in an *o*, though, may be made plural by adding *–es* (example: potato*es,* volcano*es*), and, of note, there is no known rhyme or reason for this!
- Most nouns ending in an *f* or *fe* are made plural by the *-f* or *-fe* becoming *-ves*! (example: wolf becomes *wolves*).
- Some words function as both the singular and plural form of the word (fish, deer).
- Other exceptions include *man* becomes *men, mouse* becomes *mice, goose* becomes *geese,* and *foot* becomes *feet.*

Contractions

The basic rule for making *contractions* is one area of spelling that is pretty straightforward: combine the two words by inserting an apostrophe (') in the space where a letter is omitted. For example, to combine *you* and *are*, drop the *a* and put the apostrophe in its place: *you're.*

he + is = he's
you + all = y'all (informal but often misspelled)

Note that *it's*, when spelled with an apostrophe, is always the contraction for *it is*. The possessive form of the word is written without an apostrophe as *its*.

Correcting Misspelled Words

A good place to start looking at commonly misspelled words here is with the word *misspelled*. While it looks peculiar, look at it this way: *mis* (the prefix meaning *wrongly*) + *spelled* = *misspelled*.

Let's look at some commonly misspelled words.

Commonly Misspelled Words					
accept	benign	existence	jewelry	parallel	separate
acceptable	bicycle	experience	judgment	pastime	sergeant
accidentally	brief	extraordinary	library	permissible	similar
accommodate	business	familiar	license	perseverance	supersede
accompany	calendar	February	maintenance	personnel	surprise
acknowledgement	campaign	fiery	maneuver	persuade	symmetry
acquaintance	candidate	finally	mathematics	possess	temperature
acquire	category	forehead	mattress	precede	tragedy
address	cemetery	foreign	millennium	prevalent	transferred
aesthetic	changeable	foremost	miniature	privilege	truly
aisle	committee	forfeit	mischievous	pronunciation	usage
altogether	conceive	glamorous	misspell	protein	valuable
amateur	congratulations	government	mortgage	publicly	vengeance
apparent	courtesy	grateful	necessary	questionnaire	villain
appropriate	deceive	handkerchief	neither	recede	Wednesday
arctic	desperate	harass	nickel	receive	weird
asphalt	discipline	hygiene	niece	recommend	
associate	disappoint	hypocrisy	ninety	referral	
attendance	dissatisfied	ignorance	noticeable	relevant	
auxiliary	eligible	incredible	obedience	restaurant	
available	embarrass	intelligence	occasion	rhetoric	
balloon	especially	intercede	occurrence	rhythm	
believe	exaggerate	interest	omitted	schedule	
beneficial	exceed	irresistible	operate	sentence	

Capitalization

Here's a non-exhaustive list of things that should be capitalized.

- The first word of every sentence
- The first word of every line of poetry
- The first letter of proper nouns (World War II)
- Holidays (Valentine's Day)
- The days of the week and months of the year (Tuesday, March)
- The first word, last word, and all major words in the titles of books, movies, songs, and other creative works (In the novel, *To Kill a Mockingbird*, note that *a* is lowercase since it's not a major word, but *to* is capitalized since it's the first word of the title.)
- Titles when preceding a proper noun (President Roberto Gonzales, Aunt Judy)

When simply using a word such as president or secretary, though, the word is not capitalized.

Officers of the new business must include a *president* and *treasurer*.

Seasons—spring, fall, etc.—are not capitalized.

North, *south*, *east*, and *west* are capitalized when referring to regions but are not when being used for directions. In general, if it's preceded by *the* it should be capitalized.

I'm from the South.
I drove south.

Punctuation

End Punctuation

Periods (.) are used to end a sentence that is a statement (*declarative*) or a command (*imperative*). They should not be used in a sentence that asks a question or is an exclamation. Periods are also used in abbreviations, which are shortened versions of words.

- Declarative: The boys refused to go to sleep.
- Imperative: Walk down to the bus stop.
- Abbreviations: Joan Roberts, M.D., Apple Inc., Mrs. Adamson
- If a sentence ends with an abbreviation, it is inappropriate to use two periods. It should end with a single period after the abbreviation.

The chef gathered the ingredients for the pie, which included apples, flour, sugar, etc.

Question marks (?) are used with direct questions (*interrogative*). An *indirect question* can use a period:

Interrogative: When does the next bus arrive?

Indirect Question: I wonder when the next bus arrives.

An *exclamation point (!)* is used to show strong emotion or can be used as an *interjection*. This punctuation should be used sparingly in formal writing situations.

What an amazing shot!

Whoa!

Commas

A *comma* (,) is the punctuation mark that signifies a pause—breath—between parts of a sentence. It denotes a break of flow. Proper comma usage helps readers understand the writer's intended emphasis of ideas.

In a complex sentence—one that contains a subordinate (dependent) clause or clauses—the use of a comma is dictated by where the subordinate clause is located. If the subordinate clause is located before the main clause, a comma is needed between the two clauses.

I will not pay for the steak, *because I don't have that much money*.

Generally, if the subordinate clause is placed after the main clause, no punctuation is needed.

I did well on my exam because I studied two hours the night before.

Notice how the last clause is dependent because it requires the earlier independent clauses to make sense.

Use a comma on both sides of an interrupting phrase.

I will pay for the ice cream, *chocolate and vanilla*, and then will eat it all myself.

The words forming the phrase in italics are nonessential (extra) information. To determine if a phrase is nonessential, try reading the sentence without the phrase and see if it's still coherent.

A comma is not necessary in this next sentence because no interruption—nonessential or extra information—has occurred. Read sentences aloud when uncertain.

I will pay for his chocolate and vanilla ice cream and then will eat it all myself.

If the nonessential phrase comes at the beginning of a sentence, a comma should only go at the end of the phrase. If the phrase comes at the end of a sentence, a comma should only go at the beginning of the phrase.

Other types of interruptions include the following:

- interjections: Oh no, I am not going.
- abbreviations: Barry Potter, M.D., specializes in heart disorders.
- direct addresses: Yes, Claudia, I am tired and going to bed.
- parenthetical phrases: His wife, lovely as she was, was not helpful.
- transitional phrases: Also, it is not possible.

The second comma in the following sentence is called an Oxford comma.

I will pay for ice cream, syrup, and pop.

It is a comma used after the second-to-last item in a series of three or more items. It comes before the word *or* or *and*. Not everyone uses the Oxford comma; it is optional, but many believe it is needed. The comma functions as a tool to reduce confusion in writing. So, if omitting the Oxford comma would cause confusion, then it's best to include it.

Commas are used in math to mark the place of thousands in numerals, breaking them up so they are easier to read. Other uses for commas are in dates (*March 19, 2016*), letter greetings (*Dear Sally,*), and in between cities and states (*Louisville, KY*).

Semicolons

A *semicolon (;)* is used to connect ideas in a sentence in some way. There are three main ways to use semicolons.

Link two independent clauses without the use of a coordinating conjunction:

I was late for work again; I'm definitely going to get fired.

Link two independent clauses with a transitional word:

The songs were all easy to play; therefore, he didn't need to spend too much time practicing.

Between items in a series that are already separated by commas or if necessary to separate lengthy items in a list:

Starbucks has locations in Media, PA; Swarthmore, PA; and Morton, PA.

Several classroom management issues presented in the study: the advent of a poor teacher persona in the context of voice, dress, and style; teacher follow-through from the beginning of the school year to the end; and the depth of administrative support, including ISS and OSS protocol.

Colons

A *colon* is used after an independent clause to present an explanation or draw attention to what comes next in the sentence. There are several uses.

Explanations of ideas:

They soon learned the hardest part about having a new baby: sleep deprivation.

Lists of items:

Shari picked up all the supplies she would need for the party: cups, plates, napkins, balloons, streamers, and party favors.

Time, subtitles, general salutations:

The time is 7:15.

I read a book entitled *Pluto: A Planet No More*.

To whom it may concern:

Parentheses and Dashes

Parentheses are half-round brackets that look like this: *( )*. They set off a word, phrase, or sentence that is an afterthought, explanation, or side note relevant to the surrounding text but not essential. A pair of commas is often used to set off this sort of information, but parentheses are generally used for information that would not fit well within a sentence or that the writer deems not important enough to be structurally part of the sentence.

The picture of the heart (see above) shows the major parts you should memorize.
Mount Everest is one of three mountains in the world that are over 28,000 feet high (K2 and Kanchenjunga are the other two).

See how the sentences above are complete without the parenthetical statements? In the first example, *see above* would not have fit well within the flow of the sentence. The second parenthetical statement could have been a separate sentence, but the writer deemed the information not pertinent to the topic.

The **em-dash** (—) is a mark longer than a hyphen used as a punctuation mark in sentences and to set apart a relevant thought. Even after plucking out the line separated by the dash marks, the sentence will be intact and make sense.

> Looking out the airplane window at the landmarks—Lake Clarke, Thompson Community College, and the bridge—she couldn't help but feel excited to be home.

The dashes use is similar to that of parentheses or a pair of commas. So, what's the difference? Many believe that using dashes makes the clause within them stand out while using parentheses is subtler. It's advised to not use dashes when commas could be used instead.

Ellipses

An *ellipsis* (…) consists of three handy little dots that can speak volumes on behalf of irrelevant material. Writers use them in place of words, lines, phrases, list content, or paragraphs that might just as easily have been omitted from a passage of writing. This can be done to save space or to focus only on the specifically relevant material.

> Exercise is good for some unexpected reasons. Watkins writes, "Exercise has many benefits such as…reducing cancer risk."

In the example above, the ellipsis takes the place of the other benefits of exercise that are more expected.

The ellipsis may also be used to show a pause in sentence flow.

> "I'm wondering...how this could happen," Dylan said in a soft voice.

Quotation Marks

Double quotation marks are used at the beginning and end of a direct quote. They are also used with certain titles and to indicate that a term being used is slang or referenced in the sentence. Quotation marks should not be used with an indirect quote. Single quotation marks are used to indicate a quote within a quote.

> Direct quote: "The weather is supposed to be beautiful this week," she said.
>
> Indirect quote: One of the customers asked if the sale prices were still in effect.
>
> Quote within a quote: "My little boy just said 'Mama, I want cookie,'" Maria shared.

Titles: Quotation marks should also be used to indicate titles of short works or sections of larger works, such as chapter titles. Other works that use quotation marks include poems, short stories, newspaper articles, magazine articles, web page titles, and songs.

> "The Road Not Taken" is my favorite poem by Robert Frost.
>
> "What a Wonderful World" is one of my favorite songs.

Specific or emphasized terms: Quotation marks can also be used to indicate a technical term or to set off a word that is being discussed in a sentence. Quotation marks can also indicate sarcasm.

> The new step, called "levigation," is a very difficult technique.

He said he was "hungry" multiple times, but he only ate two bites.

Use with other punctuation: The use of quotation marks with other punctuation varies, depending on the role of the ending or separating punctuation.

In American English, *periods* and *commas* always go inside the quotation marks:

"This is the last time you are allowed to leave early," his boss stated.

The newscaster said, "We have some breaking news to report."

Question marks or *exclamation points* go inside the quotation marks when they are part of a direct quote:

The doctor shouted, "Get the crash cart!"

When the question mark or exclamation point is part of the sentence, not the quote, it should be placed outside of the quotation marks:

Was it Jackie that said, "Get some potatoes at the store"?

Apostrophes

This punctuation mark, the apostrophe (') is a versatile mark. It has several different functions:

- Quotes: Apostrophes are used when a second quote is needed within a quote.

    In my letter to my friend, I wrote, "The girl had to get a new purse, and guess what Mary did? She said, 'I'd like to go with you to the store.' I knew Mary would buy it for her."

- Contractions: Another use for an apostrophe in the quote above is a contraction. *I'd* is used for *I would.*
- Possession: An apostrophe followed by the letter s shows possession (Mary's purse). If the possessive word is plural, the apostrophe generally just follows the word. Not all possessive pronouns require apostrophes.

    The trees' leaves are all over the ground.

Hyphens

The *hyphen* (-) is a small hash mark that can be used to join words to show that they are linked.

Hyphenate two words that work together as a single adjective (a compound adjective).

honey-covered biscuits

Some words always require hyphens, even if not serving as an adjective.

merry-go-round

Hyphens always go after certain prefixes like *anti-* & *all-*.

Hyphens should also be used when the absence of the hyphen would cause a strange vowel combination (*semi-engineer*) or confusion. For example, *re-collect* should be used to describe something being gathered twice rather than being written as *recollect*, which means to remember.

Expressions

Sentence Structure

Sentence Fluency

It's time to take what's been studied and put it all together in order to construct well-written sentences and paragraphs that have correct structure. Learning and utilizing the mechanics of structure will encourage effective, professional results, and adding some creativity will elevate one's writing to a higher level.

First, let's review the basic elements of sentences.

A *sentence* is a set of words that make up a grammatical unit. The words must have certain elements and be spoken or written in a specific order to constitute a complete sentence that makes sense.

1. A sentence must have a *subject* (a noun or noun phrase). The subject tells whom or what the sentence is addressing (i.e. what it is about).

2. A sentence must have an *action* or *state of being* (*a verb*). To reiterate: A verb forms the main part of the predicate of a sentence. This means that it explains what the noun is doing.

3. A sentence must convey a complete thought.

When examining writing, be mindful of grammar, structure, spelling, and patterns. Sentences can come in varying sizes and shapes; so, the point of grammatical correctness is not to stamp out creativity or diversity in writing. Rather, grammatical correctness ensures that writing will be enjoyable and clear. One of the most common methods for catching errors is to mouth the words as you read them. Many typos are fixed automatically by our brain, but mouthing the words often circumvents this instinct and helps one read what's actually on the page. Often, grammar errors are caught not by memorization of grammar rules but by the training of one's mind to know whether something *sounds* right or not.

Types of Sentences

There isn't an overabundance of absolutes in grammar, but here is one: every sentence in the English language falls into one of four categories.

- Declarative: a simple statement that ends with a period

    The price of milk per gallon is the same as the price of gasoline.

- Imperative: a command, instruction, or request that ends with a period

    Buy milk when you stop to fill up your car with gas.

- Interrogative: a question that ends with a question mark

    Will you buy the milk?

- Exclamatory: a statement or command that expresses emotions like anger, urgency, or surprise and ends with an exclamation mark

    Buy the milk now!

Declarative sentences are the most common type, probably because they are comprised of the most general content, without any of the bells and whistles that the other three types contain. They are, simply, declarations or statements of any degree of seriousness, importance, or information.

Imperative sentences often seem to be missing a subject. The subject is there, though; it is just not visible or audible because it is *implied*. Look at the imperative example sentence.

    Buy the milk when you fill up your car with gas.

You is the implied subject, the one to whom the command is issued. This is sometimes called *the understood you* because it is understood that *you* is the subject of the sentence.

Interrogative sentences—those that ask questions—are defined as such from the idea of the word *interrogation*, the action of questions being asked of suspects by investigators. Although that is serious business, interrogative sentences apply to all kinds of questions.

To exclaim is at the root of *exclamatory* sentences. These are made with strong emotions behind them. The only technical difference between a declarative or imperative sentence and an exclamatory one is the exclamation mark at the end. The example declarative and imperative sentences can both become an exclamatory one simply by putting an exclamation mark at the end of the sentences.

    The price of milk per gallon is the same as the price of gasoline!
    Buy milk when you stop to fill up your car with gas!

After all, someone might be really excited by the price of gas or milk, or they could be mad at the person that will be buying the milk! However, as stated before, exclamation marks in abundance defeat their own purpose! After a while, they begin to cause fatigue! When used only for their intended purpose, they can have their expected and desired effect.

Lengths

The ideal sentence length—the number of words in a sentence—depends upon the sentence's purpose.

It's okay for a sentence to be brief, and it's fine for a sentence to be lengthy. It's just important to make sure that long sentences do not become run-on sentences or too long to keep up with.

To keep writing interesting, vary sentence lengths, using a mixture of short, medium and long sentences.

Transitions

Transitions are the glue used to make organized thoughts adhere to one another. Transitions are the glue that helps put ideas together seamlessly, within sentences and paragraphs, between them, and (in longer documents) even between sections. Transitions may be single words, sentences, or whole paragraphs (as in the prior example). Transitions help readers to digest and understand what to feel about what has gone on and clue readers in on what is going on, what will be, and how they might react to all these factors. Transitions are like good clues left at a crime scene.

Parallel Structure in a Sentence

Parallel structure, also known as parallelism, refers to using the same grammatical form within a sentence. This is important in lists and for other components of sentences.

> Incorrect: At the recital, the boys and girls were dancing, singing, and played musical instruments.
> Correct: At the recital, the boys and girls were dancing, singing, and playing musical instruments.

Notice that in the second example, *played* is not in the same verb tense as the other verbs, nor is it compatible with the helping verb *were*. To test for parallel structure in lists, try reading each item as if it were the only item in the list.

> The boys and girls were dancing.
> The boys and girls were singing.
> The boys and girls were played musical instruments.

Suddenly, the error in the sentence becomes very clear. Here's another example:

> Incorrect: After the accident, I informed the police *that Mrs. Holmes backed* into my car, *that Mrs. Holmes got out* of her car to look at the damage, and *she was driving* off without leaving a note.
>
> Correct: After the accident, I informed the police *that Mrs. Holmes backed* into my car, *that Mrs. Holmes got out* of her car to look at the damage, and *that Mrs. Holmes drove off* without leaving a note.
>
> Correct: After the accident, I informed the police that Mrs. Holmes *backed* into my car, *got out* of her car to look at the damage, and *drove off* without leaving a note.

Note that there are two ways to fix the nonparallel structure of the first sentence. The key to parallelism is consistent structure.

Examples of Transitional Words and Phrases

Transitions have many emphases as can be seen below.

- To show emphasis: truly, in fact
- To show examples: for example, namely, specifically
- To show similarities: also, likewise
- To show dissimilarities: on the other hand, even if, in contrast
- To show progression of time: later, previously, subsequently
- To show sequence or order: next, finally
- To show cause and effect: therefore, so
- To show place or position: above, nearby, there
- To provide evidence: furthermore, then
- To summarize: finally, summarizing

Sentence Structures

A *simple sentence* has one independent clause.

I am going to win.

A *compound sentence* has two independent clauses. A conjunction—*for, and, nor, but, or, yet, so*—links them together. Note that each of the independent clauses has a subject and a verb.

I am going to win, but the odds are against me.

A *complex sentence* has one independent clause and one or more dependent clauses.

I am going to win, even though I don't deserve it.

Even though I don't deserve it is a dependent clause. It does not stand on its own. Some conjunctions that link an independent and a dependent clause are *although*, *because*, *before*, *after*, *that*, *when*, *which*, and *while*.

A *compound-complex sentence* has at least three clauses, two of which are independent and at least one that is a dependent clause.

While trying to dance, I tripped over my partner's feet, but I regained my balance quickly.

The dependent clause is *While trying to dance*.

Prewriting

Brainstorming

One of the most important steps in writing an essay is prewriting. Before drafting an essay, it's helpful to think about the topic for a moment or two, in order to gain a more solid understanding of what the task is. Then, spending about five minutes jotting down the immediate ideas that could work for the essay is recommended. It is a way to get some words on the page and offer a reference for ideas when drafting. Scratch paper is provided for writers to use any prewriting techniques such as webbing, free writing, or listing. The goal is to get ideas out of the mind and onto the page.

Considering Opposing Viewpoints

In the planning stage, it's important to consider all aspects of the topic, including different viewpoints on the subject. There are more than two ways to look at a topic, and a strong argument considers those opposing viewpoints. Considering opposing viewpoints can help writers present a fair, balanced, and informed essay that shows consideration for all readers. This approach can also strengthen an argument by recognizing and potentially refuting the opposing viewpoint(s).

Drawing from personal experience may help to support ideas. For example, if the goal for writing is a personal narrative, then the story should be from the writer's own life. Many writers find it helpful to draw from personal experience, even in an essay that is not strictly narrative. Personal anecdotes or short stories can help to illustrate a point in other types of essays as well.

Content and Organization

Moving from Brainstorming to Planning

Once the ideas are on the page, it's time to turn them into a solid plan for the essay. The best ideas from the brainstorming results can then be developed into a more formal outline. An outline typically has one main point (the thesis) and at least three sub-points that support the main point.

Here's an example:

Main Idea

- Point #1
- Point #2
- Point #3

Of course, there will be details under each point, but this approach is the best for dealing with timed writing.

Staying on Track

Basing the essay on the outline aids in both organization and coherence. The goal is to ensure that there is enough time to develop each sub-point in the essay, roughly spending an equal amount of time on each idea. Keeping an eye on the time will help. If there are fifteen minutes left to draft the essay, then it makes sense to spend about 5 minutes on each of the ideas. Staying on task is critical to success, and timing out the parts of the essay can help writers avoid feeling overwhelmed.

Parts of the Essay

The *introduction* has to do a few important things:

- Establish the *topic* of the essay in original wording (i.e., not just repeating the prompt)
- Clarify the significance/importance of the topic or purpose for writing (not too many details, a brief overview)
- Offer a *thesis statement* that identifies the writer's own viewpoint on the topic (typically one-two brief sentences as a clear, concise explanation of the main point on the topic)

Body paragraphs reflect the ideas developed in the outline. Three-four points is probably sufficient for a short essay, and they should include the following:

- A *topic sentence* that identifies the sub-point (e.g., a reason why, a way how, a cause or effect)
- A detailed *explanation* of the point, explaining why the writer thinks this point is valid
- Illustrative examples, such as personal examples or real-world examples, that support and validate the point (i.e., "prove" the point)
- A *concluding sentence* that connects the examples, reasoning, and analysis to the point being made

The *conclusion*, or final paragraph, should be brief and should reiterate the focus, clarifying why the discussion is significant or important. It is important to avoid adding specific details or new ideas to this paragraph. The purpose of the conclusion is to sum up what has been said to bring the discussion to a close.

Practice Questions

Directions for questions 1 – 35.

Select the best version of the underlined part of the sentence. If there is no mistake, choose Correct as is.

1. An important issues stemming from this meeting is that we won't have enough time to meet all of the objectives.
   a. Important issues stemming from this meeting
   b. Important issue stemming from this meeting
   c. An important issue stemming from this meeting
   d. *Correct as is*

2. The rising popularity of the clean eating movement can be attributed to the fact that experts say added sugars and chemicals in our food are to blame for the obesity epidemic.
   a. with the facts that experts say added sugars and chemicals in our food are to blame for the obesity epidemic.
   b. in the facts that experts say added sugars and chemicals in our food are to blame for the obesity epidemic.
   c. to the fact that experts saying added sugars and chemicals in our food are to blame for the obesity epidemic.
   d. *Correct as is*

3. She's looking for a suitcase that can fit all of her clothes, shoes, accessory, and makeup.
   a. clothes, shoe, accessory, and makeup.
   b. clothes, shoes, accessories, and makeup.
   c. clothes, shoes, accessories, and makeups.
   d. *Correct as is*

4. Shawn started taking guitar lessons while he wanted to become a better musician.
   a. because he wanted to become a better musician.
   b. because he wants to become a better musician.
   c. even though he wanted to become a better musician.
   d. *Correct as is*

5. Considering the recent rains we have had, it's a wonder the plants haven't drowned.
   a. Considering, the recent rains we have had, its a wonder
   b. Consider the recent rains we have had, it's a wonder
   c. Considering for how much recent rain we have had, its a wonder
   d. *Correct as is*

6. Since none of the furniture were delivered on time, we have to move in at a later date.
   a. Since all of the furniture was delivered on time
   b. Since none of the furniture was delivered on time,
   c. Since all of the furniture were delivered on time,
   d. *Correct as is*

7. It is necessary for instructors to offer tutoring to any students who need extra help in the class.
   a. for any students needing any extra help in their class.
   b. for any students that need extra help in the class.
   c. with any students who need extra help in the class.
   d. *Correct as is*

8. The fact the train set only includes four cars and one small track was a big disappointment to my son.
   a. that the train set only includes four cars and one small track were a big disappointment
   b. that the trains set only include four cars and one small track was a big disappointment
   c. that the train set only includes four cars and one small track was a big disappointment
   d. *Correct as is*

9. Because many people feel there are too many distractions to get any work done, I actually enjoy working from home.
   a. With most people
   b. While many people
   c. Maybe many people
   d. *Correct as is*

10. There were many questions about what causes the case to have gone cold, but the detective wasn't willing to discuss it with reporters.
    a. about why the case went cold
    b. about why the case is cold
    c. about what causes the case to go cold
    d. *Correct as is*

11. Every morning we would wake up, eat breakfast, and broke camp.
    a. we are waking up, eating breakfast, and breaking camp.
    b. we would wake up, eat breakfast, and break camp.
    c. would we wake up, eat breakfast, and break camp?
    d. *Correct as is*

12. Those are also memories that my siblings and me have now shared with our own children.
    a. Those are also memories that I and my siblings
    b. Those are also memories that me and my siblings
    c. Those are also memories that my siblings and I
    d. *Correct as is*

13. Ginger was born and raised in Maywood of Chicago, Illinois in 1955.
    a. in Chicago, Illinois of Maywood in 1955.
    b. in Maywood, of Chicago, Illinois in 1955.
    c. in Maywood of Chicago, Illinois, in 1955.
    d. *Correct as is*

14. Virginia was attracted to the Committee's approach toward the new legislature.
    a. Committies' approach
    b. Committees approach
    c. Committees' approach
    d. *Correct as is*

15. Franz went to a high school five minutes from his house. He played in the marching band.
    a. Franz went to a high school five minutes from his house and played in the marching band.
    b. Franz went to a high school five minutes from his house but played in the marching band.
    c. Franz went to a high school five minutes from his house, he played in the marching band.
    d. *Correct as is*

16. After school, the news team was held by a press conference that shed light on the parking lot issues.
    a. holds a press conference
    b. held a press conference
    c. holding a press conference
    d. *Correct as is*

17. In 2015; seven years later, it was finally revealed that their dog, Bear, was part Husky.
    a. In 2015. Seven years later,
    b. In 2015, seven years later,
    c. In 2015 seven years later,
    d. *Correct as is*

18. Last week, my teacher shared this information with me "One day these rules won't matter, but what's inside your head and heart will."
    a. with me. "One day these rules won't matter, but what's inside your head and heart will."
    b. with me: "One day these rules won't matter, but what's inside your head and heart will."
    c. with me: "One day these rules won't matter, but what's inside your head and heart will".
    d. *Correct as is*

19. Society raising it's children to be decent human beings with something valuable to contribute to the world.
    a. raising its children to be
    b. raises its children to be
    c. raising its' children to be
    d. *Correct as is*

20. All children can learn. Although not all children learn in the same manner.
    a. All children can learn, although not all children learn in the same manner.
    b. All children can learn although not all children learn in the same manner.
    c. All children can learn although, not all children learn in the same manner.
    d. *Correct as is*

21. If teachers set high expectations for there students, the students will rise to that high level.
    a. thare students
    b. they're students
    c. their students
    d. *Correct as is*

22. What I like about Biology class is that we focus not on the "what" of the content, but more importantly, the 'why.'
    a. but more importantly, the "why".
    b. but more importantly, the "why."
    c. but more importantly, the 'why'.
    d. *Correct as is*

23. While kids spend most of their time in school, they are dramatically shaped <u>with the influences</u> of their family.
    a. by the influences
    b. for the influences
    c. to the influences
    d. *Correct as is*

24. Because Kevin broke his leg last year, <u>he must spend</u> time in physical therapy to get ready for football season.
    a. he was spent
    b. he must to spend
    c. he must spending
    d. *Correct as is*

25. He spent much of his life helping others <u>by showing them better ways to farm, his ideas improved agricultural productivity</u> in many countries.
    a. by showing them better ways to farm; his ideas improved agricultural productivity
    b. by showing them better ways to farm his ideas improved agricultural productivity
    c. by showing them better ways to farm . . . his ideas improved agricultural productivity
    d. *Correct as is*

26. Lisa is constantly worried about being late to <u>work. This</u> is her worst nightmare.
    a. Lisa is constantly worried about being late to work: this is her worst nightmare.
    b. Lisa is constantly worried about being late to work this is her worst nightmare.
    c. Lisa is constantly worried about being late to work, this is her worst nightmare.
    d. *Correct as is*

27. When Ariana <u>goes off to college</u>, she knew she would miss her high school friends.
    a. going off to college
    b. go off to college
    c. went off to college
    d. *Correct as is*

28. <u>Although, they were closed down, they are open again after remodeling.</u>
    a. Although they were closed down, they are open again after remodeling.
    b. Although they were closed down they are open again after remodeling.
    c. Although they were closed down. They are open again after remodeling.
    d. *Correct as is*

29. Ms. Richie <u>told her second period class they were</u> her favorite.
    a. tells her second period class they were
    b. told her second period class they are
    c. told her second period class they are being
    d. *Correct as is*

30. <u>'Wait,' he said, 'for the bus to arrive.'</u>
    a. "Wait," he said, "for the bus to arrive."
    b. "Wait", he said, "for the bus to arrive".
    c. "Wait" he said "for the bus to arrive."
    d. *Correct as is*

31. Next time I go to the <u>beach; I</u> hope the sun comes out.
   a. Next time I go to the beach: I hope the sun comes out.
   b. Next time I go to the beach, I hope the sun comes out.
   c. Next time I go to the beach. I hope the sun comes out.
   d. *Correct as is*

32. No one was <u>best than</u> Teresa at climbing a tree.
   a. No one was as best as Teresa at climbing a tree.
   b. No one was more better than Teresa at climbing a tree.
   c. No one was better than Teresa at climbing a tree.
   d. *Correct as is*

33. She called the doctor's office eight times <u>before she reached</u> Dr. Yezzi.
   a. after she got reach of
   b. before she reaches
   c. because she had gotten reach of
   d. *Correct as is*

34. If you're going to the dance on Saturday, <u>can we drive your red mustang.</u>
   a. can we be driving your red mustang?
   b. can we drive your red mustang?
   c. can we drive your red mustang!
   d. *Correct as is*

35. The worst part about the surgery is <u>that, I</u> won't be able to eat for twenty-four hours the day before.
   a. The worst part about the surgery is that I won't be able to eat for twenty-four hours the day before.
   b. The worst part about the surgery is that: I won't be able to eat for twenty-four hours the day before.
   c. The worst part about the surgery is that; I won't be able to eat for twenty-four hours the day before.
   d. *Correct as is*

Directions for questions 36 – 48

Read the whole sentence below. There might be a mistake in sentence structure. Choose the answer that is written most clearly. If there is no error, choose Correct as is.

36. You must take and edit, in order to become a famous photographer, pictures every single day.
   a. Every single day, in order to become a famous photographer, you must take and edit pictures.
   b. A famous photographer, in order to become you must take and edit pictures every single day.
   c. In order to become a famous photographer, you must take and edit pictures every single day.
   d. *Correct as is*

37. I went to the café to get a coffee, a sandwich, and some chips, while I was on my lunch break.
   a. A coffee, a sandwich, and some chips I got while I was on my lunch break when I went to the café.
   b. To get coffee, a sandwich, and some chips, I went to the café while on my lunch break.
   c. While I was on my lunch break, I went to the café to get a coffee, a sandwich, and some chips.
   d. *Correct as is*

38. Rachel sold essential oils to everyone in her family, except her Aunt Lydia, who was allergic to lavender.
    a. Except her Aunt Lydia, who was allergic to lavender, Rachel sold essential oils to everyone in her family.
    b. Rachel sold essential oils to everyone, except her Aunt Lydia, in her family, who was allergic to lavender.
    c. To everyone in her family Rachel sold essential oils, except her Aunt Lydia, who was allergic to lavender.
    d. *Correct as is*

39. For pizza and ice cream the team went out, after they won the championship.
    a. Winning the championship, afterwards the whole team went out for pizza and ice cream.
    b. After they won the championship, the whole team went out for pizza and ice cream.
    c. They won the championship for which after the team went out for pizza and ice cream.
    d. *Correct as is*

40. Bob wanted to leave for vacation on Friday, and Anna wanted to leave for vacation on Saturday, because he had the day off and she was too tired from the work week.
    a. Bob wanted to leave for vacation on Friday because he had the day off, but Anna wanted to leave for vacation on Saturday because she was too tired from the work week.
    b. Bob wanting to leave for vacation on Friday because he had the day off, while Anna wanting to leave for vacation on Saturday because she was too tired from the work week.
    c. Bob because he had the day off, and Anna because she was too tired from the work week, one wanted to leave on Friday and one wanted to leave on Saturday.
    d. *Correct as is*

41. Having arrived late to the hospital, it was clear that Jackson's mom forgot to pick him up after his surgery.
    a. Jackson's mom was late to the hospital, forgetting to pick Jackson up after his surgery, because she forgot.
    b. Jackson's mom clearly forgot to pick him up after his surgery because she arrived late to the hospital.
    c. Jackson's mom forgetting to pick him up after his surgery arrived late to the hospital.
    d. *Correct as is*

42. Josefina, she lives in New York and can't wait to find a new studio there being she dislikes the place she lives in now.
    a. Although she lives in New York, Josefina can't wait to find a new studio, disliking the place she lives in now.
    b. Disliking the place she lives in now, Josefina lives in New York and can't wait to find another studio.
    c. Josefina lives in New York and can't wait to find a new studio there because she dislikes the place she lives in now.
    d. *Correct as is*

43. Although Mark majored in Chemistry, he became a writer and won a Pulitzer Prize.
   a. Mark became a writer and won a Pulitzer Prize because he majored in Chemistry.
   b. Winning a Pulitzer Prize and becoming a writer, Mark majored in Chemistry.
   c. Mark, majoring in Chemistry, winning a Pulitzer Prize, and becoming a writer.
   d. *Correct as is*

44. In order to fund his college career, Brendan takes out student loans and worked two jobs in a nearby city.
   a. His college career being funded, Brendan works two jobs and takes out student loans.
   b. Brendan works two jobs and takes out student loans, funding his college career.
   c. In order to fund his college career, Brendan takes out student loans and works two jobs in a nearby city.
   d. *Correct as is*

45. Babysitting her nephews was not Brittany's ideal job; however, she looked forward to seeing them every weekend.
   a. However, always looking forward to seeing them, Brittany babysat her nephews every weekend but it was not her ideal job.
   b. Brittany was always looking forward to the weekend, babysitting her nephews, although not being her ideal job.
   c. She always looked forward to babysitting her nephews every weekend; although this wasn't Brittany's ideal job.
   d. *Correct as is*

46. Jack enjoyed the warm skies and the trips to the beach; always rejuvenated him in summertime.
   a. In summertime, Jack was always rejuvenated, he enjoyed the warm skies, the trips to the beach.
   b. Jack enjoyed summertime—the warm skies and the trips to the beach always rejuvenated him.
   c. Always rejuvenating him, the warm skies and the trips to the beach in summertime.
   d. *Correct as is*

47. Geology or Space Science Jodi could not decide which subject she liked better.
   a. Which subject she liked better, Geology or Space Science, Jodi could not decide.
   b. Jodi could not decide which subject she liked better: Geology or Space Science.
   c. Which subject (Geology or Space Science) Jodi liked better she could not decide.
   d. *Correct as is*

48. The tension she had been holding was released, and she was able to breathe again, after her presentation.
   a. After she gave her presentation, the tension she had been holding was released, and she was able to breathe again.
   b. She was able to breathe again—after her presentation—the tension she had been holding again was released.
   c. The tension she had been holding, after her presentation, was released, and she was able to breathe again.
   d. *Correct as is*

Answer Explanations

1. C: In this answer, the article and subject agree, and the subject and predicate agree. Choice *A* is incorrect because the plural subject *issues* does not agree with the singular verb *is*. Choice *B* is incorrect because an article is needed before *important issue*. Choice *D* is incorrect because the article (*an*) and the noun (*issues*) do not agree in number.

2. D: Choices *A* and *B* both use the expression *attributed to the fact* incorrectly. It can only be attributed *to* the fact, not *with* or *in* the fact. Choice *C* incorrectly uses a gerund, *saying*, when it should use the present tense of the verb *say*.

3. B: Choice *B* is correct because it uses correct parallel structure of plural nouns. Choice *A* is incorrect because it again uses the singular *accessory*, and it uses the singular *shoe*. Choice *C* is incorrect because it pluralizes *makeup*, which is already in plural form. Choice *D* is incorrect because the word *accessory* is in singular form.

4. A: In a cause/effect relationship, it is correct to use the word because in the clausal part of the sentence. This can eliminate both Choices *C and D* which don't clearly show the cause/effect relationship. Choice *B* is incorrect because it uses the present tense, when the first part of the sentence is in the past tense. It makes grammatical sense for both parts of the sentence to be in present tense.

5. D: In Choice *B*, the present tense form of the verb *consider* creates an independent clause joined to another independent clause with only a comma, which is a comma splice and grammatically incorrect. Both *A* and *C* use the possessive form of *its*, when it should be the contraction *it's* for *it is*. Choice *A* also includes incorrect comma placement.

6. B: Choice *A* uses *all* again, and is missing the comma, which is necessary to set the dependent clause off from the independent clause. Choice *C* also uses the wrong verb form and uses the word *all* in place of *none*, which doesn't make sense in the context of the sentence. Choice *D* uses the plural form of the verb, when the subject is the pronoun *none*, which needs a singular verb.

7. D: Answer Choice *D* uses the best, most concise word choice. Choice *A* uses the preposition *for* and the additional word *any*, making the sentence wordy and less clear. Choice *B* uses the pronoun *that* to refer to people instead of *who*. *C* incorrectly uses the preposition *with*.

8. C: Choice *A* changes the verb to *were*, which is in plural form and does not agree with the singular subject. Choice *B* pluralizes *trains* and uses the singular form of the word *include*, so it does not agree with the word *set*. Choice *D* is missing the word *that*, which is necessary for the sentence to make sense.

9. B: Choice *B* uses the best choice of words to create a subordinate and independent clause. Choice *A* uses *with*, which does not make grammatical sense. In *C*, the word *maybe* creates two independent clauses, which are not joined properly with a comma. In Choice *D*, *because* makes it seem like this is the reason they enjoy working from home, which is incorrect.

10. A: Choices *C* and *D* use additional words and phrases that are not necessary. Choice *B* is more concise, but uses the present tense of *is*. This does not agree with the rest of the sentence, which uses past tense. The best choice is Choice *A*, which uses the most concise sentence structure and is grammatically correct.

11. B: This sentence calls for parallel structure. Choice *B* is correct because the verbs "wake," "eat," and "break" are consistent in tense and parts of speech. Choice *A* is incorrect because it breaks tense with the rest of the passage. "Waking," "eating," and "breaking" are all present participles, and the context around the sentence is in past tense. Choice *C* is incorrect because this turns the sentence into a question, which doesn't make sense within the context. Choice *D* is incorrect because the words "wake" and "eat" are present tense while the word "broke" is in past tense.

12. C: The rules for "me" and "I" is that one should use "I" when it is the subject pronoun of a sentence, and "me" when it is the object pronoun of the sentence. Break the sentence up to see if "I" or "me" should be used. To say "Those are memories that I have now shared" is correct, rather than "Those are memories that me have now shared." Choice *A* is incorrect because "my siblings" should come before "I."

13. D: Choice *D* is correct because there should be a comma between the city and state. Choice *A* is incorrect because the order of the sentence designates that Chicago, Illinois is in Maywood, which is incorrect. Choice *B* is incorrect because the comma after "Maywood" interrupts the phrase "Maywood of Chicago." Choice *C* is incorrect because a comma after "Illinois" is unnecessary.

14. D: Choice *D* is correct because the Committee is one entity, therefore the possession should show the "Committee's approach" with the apostrophe between the "y" and the "s." Choice *A* is incorrect because the word "Committies" is spelled wrong. Choice *B* is incorrect because the word "Committees" should not be plural and should have an apostrophe to show possession. Choice *C* is incorrect because the apostrophe indicates that the word "Committees" is plural.

15. A: Choice *A* is correct because the conjunction "and" is the best way to combine the two independent clauses. Choice *B* is incorrect because the conjunction "but" indicates a contrast, and there is no contrast between the two clauses. Choice *C* is incorrect because the introduction of the comma after "house" with no conjunction creates a comma splice. Choice *D* is incorrect because the word "he" becomes repetitive since the two clauses can be joined together.

16. B: Choice *B* is correct because it provides the correct verb tense and also makes sense within the context of the passage. Choice *A* is incorrect because the verb tense is inconsistent with the rest of the sentence. Choice *C* is incorrect because, with this use of the sentence, it would create a fragment because the verb "holding" has no helping verb in front of it. Choice *D* is incorrect because it changes the meaning of the sentence.

17. B: Choice *B* is correct. Choice *A* is incorrect because the sentence "In 2015." is a fragment. Choice *C* is incorrect because there should be a comma after introductory phrases in general, such as "In 2015," and Choice *C* omits a comma. Choice *D* is incorrect because there should be an independent clause on either side of a semicolon, and the phrase "In 2015" is not an independent clause.

18. B: Choice *B* is correct. Here, a colon is used to introduce an explanation. Colons either introduce explanations or lists. Additionally, the quote ends with the punctuation inside the quotes, unlike Choice *C*.

19. B: The word "raising" in Choice *A* makes the sentence grammatically incorrect. Choice *C* adds an apostrophe at the end of "its." While adding an apostrophe to most words would indicate possession, adding *'s* to the word "it" indicates a contraction. The possessive form of the word "it" is "its." The contraction "it's" denotes "it is." Thus, Choice *D* is incorrect.

20. A: This sentence must have a comma before "although" because the word "although" is connecting two independent clauses. Thus, Choices *B* and *C* are incorrect. Choice *D* is incorrect because the second sentence in the underlined section is a fragment.

21. C: Choice *C* is the correct choice because the word "their" indicates possession, and the text is talking about "their students," or the students of someone. Choice *A* is not a word. Choice *B*, "they're," is a contraction and means "they are." Choice *D*, "there," means at a certain place and is incorrect.

22. B: Choice *B* uses all punctuation correctly in this sentence. Punctuation here should go inside the quotes, making Choice *A* incorrect. In American English, single quotes should only be used if they are quotes within a quote, making Choices *C* and *D* incorrect.

23. A: The correct choice for this sentence is that "they are . . . shaped by the influences." The prepositions "for," "to," and "with" do not make sense in this context. People are *shaped by*, not *shaped for, shaped to,* or *shaped with*.

24. D: Choice *D* is the best choice for this sentence. Choice *A* is incorrect because adding "was" to "spent" is incorrect verb usage in the sentence. Choice *B* is incorrect. Adding "to spend" to "he must" is grammatically incorrect. Choice *C* is incorrect because saying "he must spending" is also incorrect usage of the verb "spending."

25. A: Out of these choices, a semicolon would be the best fit because there is an independent clause on either side of the semicolon, and the two sentences closely relate to each other. Choice *B* is incorrect; omitting punctuation here creates a run-on sentence. Choice *C* is incorrect because an ellipsis (. . .) is used to designate an omission in the text. Choice *D* is incorrect because putting a comma between two independent clauses (i.e. complete sentences) creates a comma splice.

26. D: There should be no change here. Both underlined sentences are complete and do not need changing. Choice *A* is incorrect. The underlined portion could *possibly* act with a colon. However, it's not the best choice, so omit Choice *A*. Choice B is incorrect because there is no punctuation between the two independent clauses; it is considered a run-on. Choice *C* is incorrect because placing a comma between two independent clauses creates a comma splice.

27. C: Choice *C* is correct because it uses the appropriate verb tense and usage. Choice *A* is incorrect. "When Ariana going off to college" is incorrect usage of the verb "to go." Choice *B* is incorrect because of the faulty subject/verb agreement. The verb "go" should be paired with a plural subject, and "Ariana" is singular. Choice *D* is incorrect. The verb agrees with the subject; however, the rest of the sentence is in past tense, so this verb should also be in past tense.

28. A: Choice *A* is correct because it connects a dependent clause (Although they were closed down) with an independent clause (they are open again after remodeling) with a comma. Choice *B* is a run-on sentence. Choice *C* creates a fragment beginning with "Although." Choice *D* puts a comma after the word "Although," which is incorrect.

29. D: The sentence is correct as is. Choice *A* combines a present tense verb (tells) with a past tense verb (were), making it incorrect. Choice *B* combines a past tense verb (told) with a present tense verb (are), making it incorrect. Choice *C* combines a past tense verb (told) with a present participle (are being), which is incorrect. Make sure the verb tense within the sentence is consistent.

30. A: The correct choice is *A*. Choice *B* places punctuation outside the quotation marks, and punctuation should be inside the quotation marks. Choice *C* has no punctuation at all, which is incorrect. Choice *D* uses single quotes, which is incorrect in standard American English when using dialogue.

31. B: Choice *A* is incorrect. It uses a colon instead of a comma. Choices *C* and *D* are also incorrect. For both a semicolon and a period, there must be a complete sentence on either side of the punctuation mark. The statement "Next time I go to the beach" is not a complete sentence. Therefore, the sentence calls for a comma.

32. C: No one was better than Teresa at climbing a tree. Choice *A* is incorrect; we could say "as good as" but "as best as" is not the correct usage. Choice *B* is also incorrect; adding "more" to "better" is repetitive. Choice *D* is incorrect because "best than" is incorrect usage.

33. D: Choice *D* is the best answer, correct as is. Choice *A* doesn't make sense in context. She wouldn't call the doctor's office eight times *after* she reached Dr. Yezzi, but before. Also, "she got reach" is more awkward than "she reached" in the context. Choice *B* is incorrect because the original sentence is in past tense, so the choice is inconsistent with the tense. Choice *C* is incorrect; she wouldn't have called the doctor's office *because she had gotten reach* of Dr. Yezzi. This choice does not make sense in this context.

34. B: Choice *B* has correct language usage as well as correct punctuation. Choice *A* is incorrect because "can we be driving" is not correct verb usage. Choice *C* is incorrect because the sentence needs a question mark, not an exclamation mark. Choice *D* is incorrect because the sentence needs a question mark, not a period.

35. A: The sentence does not need punctuation between the words "that" and "I." Choice *B* is incorrect. If "that" were changed to "this," then the sentence would make sense with a colon; however, as-is, the sentence does not make sense. Choice *C* is incorrect. A semicolon needs a complete sentence on either side of it, and "The worst part about the surgery is that" is not a complete sentence. Choice *D* is incorrect. The comma interrupts the sentence.

36. C: In order to become a famous photographer, you must take and edit pictures every single day. This is the clearest form of the sentence. It consists of a dependent clause followed by an independent clause, separated by a comma. The other choices break the sentence up into compartmental language that is difficult to comprehend.

37. C: While I was on my lunch break, I went to the café to get a coffee, a sandwich, and some chips. This sentence is the easiest to comprehend. Choice *A* is awkwardly worded. Choice *B* has a dangling modifier at the beginning of the sentence, so it is incorrect. Choice *D* comes close because it is grammatically correct. However, it is better to start with "While I was on my lunch break" because the sentence then proceeds in chronological order.

38. D: The best version of the sentence is Choice *D*, "Rachel sold essential oils to everyone in her family, except her Aunt Lydia, who was allergic to lavender." The rest of the choices invert the phrases in an awkward way. In this sentence, it is best to begin with an independent clause "Rachel sold essential oils to everyone in her family" so that the dependent clauses don't act as random chunks at the beginning of the sentence.

39. B: After they won the championship, the whole team went out for pizza and ice cream. This is the most organized version of this sentence. Choice *A* begins with a dangling modifier, which is incorrect. Choice *C's* word choice is awkward, especially with the "for which after" phrase. Choice *D* has an awkward inversion. A solid sentence structure will usually begin with a subject and a verb, with the preposition to follow.

40. A: The best version of the sentence is Choice *A*. Choice *B* is incorrect. For correct usage, the verb "wanting" should be "wanted." Choice *C* is incorrect because it is not clear which "one" wanted to leave on which day, given the current organization of the sentence. Choice *D* is incorrect. Although we know which one wanted to leave which day because of the pronouns, the sentence is awkward the way it is written.

41. B: Choice *B* gives the most straightforward sentence out of the rest of the answer choices. Choice *A* is incorrect because it uses the same subject, "Jackson's mom," with the verbs "was late," "forgetting," and "she forgot," using faulty parallel structure as well as repetition. Choice *C* is incorrect. It should set the phrase "forgetting to pick him up after his surgery" off with commas. However, even then, it is not the clearest version of the sentence. Choice *D* is incorrect because it has a dangling modifier. As it is, the subject "it" arrived late to the hospital, not Jackson's mom.

42. C: Choice *C* is correct. Choice *A* uses "Although," which doesn't make much sense in the context of the sentence. "Although" usually indicates a contradiction of some kind, and there is no contradiction here. Choice *B* is incorrect. Although grammatically correct, Choice *B* isn't as straightforward as Choice *C*. Choice *D* is incorrect because it uses duplicate subjects, "Josefina" and "she."

43. D: Choice *D* is correct because it gives us the most logical version of the sentence. Choice *A* is incorrect because the conjunction "because" does not fit in the context of the sentence. Choice *B* is incorrect because it doesn't give us the connection of what's important about Mark majoring in Chemistry and becoming a writer like Choice *D* does. Choice *D* shows the irony in the two situations by giving us the word "Although." Choice *C* is incorrect. The words that begin with "-ing" are present participles, and the verb phrase functions as an adverb, and not a verb. Therefore, this sentence is missing a verb to its subject, "Mark."

44. C: The best choice is *C*; the clause that begins "in order to" is an effective way to begin this sentence because it shows the purpose of Brendan working and taking out loans. Choices *A* and *B* aren't as clear without the clause beginning with "in order to," and Choice *A* uses passive voice ineffectively. Choice *D* uses "takes out" and "worked" for its verbs, and therefore does not maintain parallel structure.

45. D: This sentence is correct-as-is. Choice *A* is incorrect because the sentence begins with "however," yet fails to make a contrast. Beginning with "however" in this way is confusing. Choice *B* is incorrect because the dependent clause, "although not being her ideal job" is not grammatically consistent with "babysitting her nephews." Even if it were consistent, the sentence is awkward with the interrupting phrases. Choice *C* is incorrect because the dependent clause "although this wasn't Brittany's ideal job" should be separated with a comma, not a semicolon. Semicolons must have complete sentences on either side of them.

46. B: Choice *B* is the clearest form of this sentence. Choice *A* is incorrect; if there were a period or semicolon after "rejuvenated," this choice would be a possibility. However, this creates a comma splice, and the rest of the commas are excessive. Choice *C* is incorrect because it is not a complete sentence. Choice *D* is incorrect. Again, semicolons must have complete sentences on either side of them, and "always rejuvenated him in summertime" is not a complete sentence.

47. B: Choice *B* is correct because it is the most straightforward version of this sentence, and it uses the colon correctly. Choice *A* is incorrect. This sentence is inverted. It is clearer if the subject "Jodi" comes first. Choice *C* is incorrect. This sentence is also inverted, and although it could possibly be used as the correct choice, it is not the best choice.

48. A: Choice *A* is the best answer here. Choice *B* is incorrect. Em-dashes can be used in place of commas. However, it is uncertain whether "after her presentation" is a clause attributed to the first part of the sentence or the last part of the sentence. Choice *C* is incorrect because it has too many interruptions. Choice *A* is much clearer. Finally, Choice *D* is incorrect. This answer is possible—however, the second comma is not needed, and Choice *A* tells the event as it happened in chronological order.

Writing

On the CHSPE, you will be presented with one persuasive writing task. For the essay, you will be asked to write a passage concerning your view over an issue. Keep in mind a specific audience, and be consistent in addressing that audience while you write. Back up your argument with evidence and logical reasoning.

The essay scoring scale goes from a 5 (most effective) to a 1 (least effective). Below are some points that will help you receive a score of 5:

- Clearly state your argument with appropriate reasoning
- Organize the essay well
- Be aware of the audience and their concerns
- Variate the sentence structure
- Create clear and precise word choice
- Reread for errors in grammar or usage

Remember the conventions of prewriting and apply them to writing your essay. Here is a checklist to help ensure you are including everything in your essay:

- Write about the topic
- Express ideas in complete sentences
- Include sufficient details to support ideas
- Ensure Nothing is off-topic or irrelevant
- Organize ideas in a logical way
- Assign each paragraph a topic sentence
- Capitalize the appropriate words
- Use correct punctuation
- Spell words correctly
- Make sure handwriting is clear

Writing Task

Prepare an essay of about 300 – 600 words on the topic below.

Some people feel that sharing their lives on social media sites such as Facebook, Instagram, and Snapchat is fine. They share every aspect of their lives, including pictures of themselves and their families, what they ate for lunch, who they are dating, and when they are going on vacation. They even say that if it's not on social media, it didn't happen. Other people believe that sharing so much personal information is an invasion of privacy and could prove dangerous. They think sharing personal pictures and details invites predators, cyberbullying, and identity theft.

Write an essay to someone who is considering whether to participate in social media. Take a side on the issue and argue whether or not he/she should join a social media network. Use specific examples to support your argument.

Math

Number Sense and Operations

Addition with Whole Numbers and Fractions

Addition combines two quantities together. With whole numbers, this is taking two sets of things and merging them into one, then counting the result. For example, 4 + 3 = 7. When adding numbers, the order does not matter: 3 + 4 = 7, also. Longer lists of whole numbers can also be added together. The result of adding numbers is called the *sum*.

With fractions, the number on top is the *numerator*, and the number on the bottom is the *denominator*. To add fractions, the denominator must be the same—a *common denominator*. To find a common denominator, the existing numbers on the bottom must be considered, and the lowest number they will both multiply into must be determined. Consider the following equation:

$$\frac{1}{3}+\frac{5}{6}=?$$

The numbers 3 and 6 both multiply into 6. Three can be multiplied by 2, and 6 can be multiplied by 1. The top and bottom of each fraction must be multiplied by the same number. Then, the numerators are added together to get a new numerator. The following equation is the result:

$$\frac{1}{3}+\frac{5}{6}=\frac{2}{6}+\frac{5}{6}=\frac{7}{6}$$

Subtraction with Whole Numbers and Fractions

Subtraction is taking one quantity away from another, so it is the opposite of addition. The expression 4 – 3 means taking 3 away from 4. So, 4 – 3 = 1. In this case, the order matters, since it entails taking one quantity away from the other, rather than just putting two quantities together. The result of subtraction is also called the *difference*.

To subtract fractions, the denominator must be the same. Then, subtract the numerators together to get a new numerator. Here is an example:

$$\frac{1}{3}-\frac{5}{6}=\frac{2}{6}-\frac{5}{6}=\frac{-3}{6}=-\frac{1}{2}$$

Multiplication with Whole Numbers and Fractions

Multiplication is a kind of repeated addition. The expression 4×5 is taking four sets, each of them having five things in them, and putting them all together. That means $4 \times 5 = 5 + 5 + 5 + 5 = 20$. As with addition, the order of the numbers does not matter. The result of a multiplication problem is called the *product.*

To multiply fractions, the numerators are multiplied to get the new numerator, and the denominators are multiplied to get the new denominator:

$$\frac{1}{3} \times \frac{5}{6} = \frac{1 \times 5}{3 \times 6} = \frac{5}{18}$$

When multiplying fractions, common factors can *cancel* or *divide into one another*, when factors that appear in the numerator of one fraction and the denominator of the other fraction. Here is an example:

$$\frac{1}{3} \times \frac{9}{8} = \frac{1}{1} \times \frac{3}{8} = 1 \times \frac{3}{8} = \frac{3}{8}$$

The numbers 3 and 9 have a common factor of 3, so that factor can be divided out.

Division with Whole Numbers and Fractions

Division is the opposite of multiplication. With whole numbers, it means splitting up one number into sets of equal size. For example, $16 \div 8$ is the number of sets of eight things that can be made out of sixteen things. Thus, $16 \div 8 = 2$. As with subtraction, the order of the numbers will make a difference, here. The answer to a division problem is called the *quotient*, while the number in front of the division sign is called the *dividend*, and the number behind the division sign is called the *divisor*.

To divide fractions, the first fraction must be multiplied with the reciprocal of the second fraction. The *reciprocal* of the fraction $\frac{x}{y}$ is the fraction $\frac{y}{x}$. Here is an example:

$$\frac{1}{3} \div \frac{5}{6} = \frac{1}{3} \times \frac{6}{5} = \frac{6}{15} = \frac{2}{5}$$

Recognition of Decimals

The *decimal system* is a way of writing out numbers that uses ten different numerals: 0, 1, 2, 3, 4, 5, 6, 7, 8, and 9. This is also called a "base ten" or "base 10" system. Other bases are also used. For example, computers work with a base of 2. This means they only use the numerals 0 and 1.

The *decimal place* denotes how far to the right of the decimal point a numeral is. The first digit to the right of the decimal point is in the *tenths* place. The next is the *hundredths*. The third is the *thousandths*.

So, 3.142 has a 1 in the tenths place, a 4 in the hundredths place, and a 2 in the thousandths place.

The *decimal point* is a period used to separate the *ones* place from the *tenths* place when writing out a number as a decimal.

A *decimal number* is a number written out with a decimal point instead of as a fraction, for example, 1.25 instead of $\frac{5}{4}$. Depending on the situation, it can sometimes be easier to work with fractions and sometimes easier to work with decimal numbers.

A decimal number is *terminating* if it stops at some point. It is called *repeating* if it never stops but repeats a pattern over and over. It is important to note that every rational number can be written as a terminating decimal or as a repeating decimal.

Addition with Decimals

To add decimal numbers, each number needs to be lined up by the decimal point in vertical columns. For each number being added, the zeros to the right of the last number need to be filled in so that each of the numbers has the same number of places to the right of the decimal. Then, the columns can be added together. Here is an example of 2.45 + 1.3 + 8.891 written in column form:

$$\begin{array}{r} 2.450 \\ 1.300 \\ \underline{+\ 8.891} \end{array}$$

Zeros have been added in the columns so that each number has the same number of places to the right of the decimal.

Added together, the correct answer is 12.641:

$$\begin{array}{r} 2.450 \\ 1.300 \\ \underline{+\ 8.891} \\ 12.641 \end{array}$$

Subtraction with Decimals

Subtracting decimal numbers is the same process as adding decimals. Here is 7.89 – 4.235 written in column form:

$$\begin{array}{r} 7.890 \\ \underline{-\ 4.235} \\ 3.655 \end{array}$$

A zero has been added in the column so that each number has the same number of places to the right of the decimal.

Multiplication with Decimals

The simplest way to multiply decimals is to calculate the product as if the decimals are not there, then count the number of decimal places in the original problem. Use that total to place the decimal the same number of places over in your answer, counting from right to left. For example, 0.5 x 1.25 can be rewritten and multiplied as 5 x 125, which equals 625. Then the decimal is added three places from the right for .625.

The final answer will have the same number of decimal *points* as the total number of decimal *places* in the problem. The first number has one decimal place, and the second number has two decimal places. Therefore, the final answer will contain three decimal places:

$$0.5 \times 1.25 = 0.625$$

Division with Decimals

Dividing a decimal by a whole number entails using long division first by ignoring the decimal point. Then, the decimal point is moved the number of places given in the problem.

For example, $6.8 \div 4$ can be rewritten as $68 \div 4$, which is 17. There is one non-zero integer to the right of the decimal point, so the final solution would have one decimal place to the right of the solution. In this case, the solution is 1.7.

Dividing a decimal by another decimal requires changing the divisor to a whole number by moving its decimal point. The decimal place of the dividend should be moved by the same number of places as the divisor. Then, the problem is the same as dividing a decimal by a whole number.

For example, $5.72 \div 1.1$ has a divisor with one decimal point in the denominator. The expression can be rewritten as $57.2 \div 11$ by moving each number one decimal place to the right to eliminate the decimal. The long division can be completed as $572 \div 11$ with a result of 52. Since there is one non-zero integer to the right of the decimal point in the problem, the final solution is 5.2.

In another example, $8 \div 0.16$ has a divisor with two decimal points in the denominator. The expression can be rewritten as $800 \div 16$ by moving each number two decimal places to the right to eliminate the decimal in the divisor. The long division can be completed with a result of 50.

Percents

The word *percent* comes from the Latin phrase for "per one hundred." A *percent* is a way of writing out a fraction. It is a fraction with a denominator of 100. Thus, $65\% = \frac{65}{100}$.

To convert a fraction to a percent, the denominator is written as 100. For example, $\frac{3}{5} = \frac{60}{100} = 60\%$.

In converting a percent to a fraction, the percent is written with a denominator of 100, and the result is simplified. For example, $30\% = \frac{30}{100} = \frac{3}{10}$.

Percent Problems

The basic percent equation is the following:

$$\frac{is}{of} = \frac{\%}{100}$$

The placement of numbers in the equation depends on what the question asks.

Example 1
Find 40% of 80.

Basically, the problem is asking, "What is 40% of 80?" The 40% is the percent, and 80 is the number to find the percent "of." The equation is:

$$\frac{x}{80} = \frac{40}{100}$$

After cross-multiplying, the problem becomes 100x = 80(40). Solving for x gives the answer: x = 32.

Example 2

What percent of 100 is 20?

The 20 fills in the "is" portion, while 100 fills in the "of." The question asks for the percent, so that will be x, the unknown. The following equation is set up:

$$\frac{20}{100} = \frac{x}{100}$$

Cross-multiplying yields the equation 100x = 20(100). Solving for x gives the answer of 20%.

Example 3

30% of what number is 30?

The following equation uses the clues and numbers in the problem:

$$\frac{30}{x} = \frac{30}{100}$$

Cross-multiplying results in the equation 30(100) = 30x. Solving for x gives the answer x = 100.

Estimating

Estimation is finding a value that is close to a solution but is not the exact answer. For example, if there are values in the thousands to be multiplied, then each value can be estimated to the nearest thousand and the calculation performed. This value provides an approximate solution that can be determined very quickly.

Ordering of Numbers

In counting, when a number appears after another number in order, that number will be one more. On the other hand, when a number appears before another number in order, that number will be one less. This idea is useful when counting backward. Also, zero means that there is none of something. This idea can be seen by taking away all of something so that there are zero items left. Also, learning to count by tens starting at any number is a key concept. Once a new number is learned, learning how to read and write that number is also important.

Placing numbers in an order in which they are listed from smallest to largest is known as *ordering*. When items are listed by using numbers in order, the *ordinal numbers*, 1st, 2nd, 3rd, 4th, ..., can be used.

When you order numbers the right way, you can more easily compare the different amounts of items. When you compare numbers, you show whether two amounts are the same or different. Teachers can show two different quantities of items in the classroom. Then they can discuss which amount is lesser or greater. This exercise also can be used in order to classify numbers from the smallest amount to the largest amount.

Being able to compare any two whole numbers without a visual representation is also an important task. Each whole number relates to a certain amount. This amount can be ranked and compared to other amounts. Knowing the right vocabulary relating to ordering and comparing is important. The *equals sign* is $=$. It shows that two numbers are the same on either side of the symbol. For example, $28 = 28$. The symbols that are used for comparison are < to represent *less than*, > to represent *greater than*. The symbols ≤ to represent *less than or equal to*, and ≥ to represent *greater than or equal to*, and ≠ to represent *not equal to* can also be used.

You can compare numbers with any number of digits when you use these symbols. For example, the expression $77 < 100$, should be understood as 77 is less than 100. The expression $44 > 23$ should be understood as 44 is greater than 23. The expression $22 \neq 24$ should be understood as 22 is not equal *to 24*. Also, both $36 = 36$ and $36 \leq 36$ can be written because both "36 equals 36" and "36 is less than or equal to 36" applies.

Properties of Exponents

Exponents are used in mathematics to express a number or variable multiplied by itself a certain number of times. For example, x^3 means *x* is multiplied by itself three times. In this expression, x is called the *base*, and 3 is the *exponent*. Exponents can be used in more complex problems when they contain fractions and negative numbers.

Fractional exponents can be explained by looking first at the inverse of exponents, which are *roots*. Given the expression x^2, the square root can be taken, $\sqrt{x^2}$, cancelling out the 2 and leaving *x* by itself, if *x* is positive. Cancellation occurs because $\sqrt{x}$ can be written with exponents, instead of roots, as $x^{\frac{1}{2}}$. The numerator of 1 is the exponent, and the denominator of 2 is called the root (which is why it's referred to as *square root*). Taking the square root of x^2 is the same as raising it to the $\frac{1}{2}$ power. Written out in mathematical form, it takes the following progression:

$$\sqrt{x^2} = (x^2)^{\frac{1}{2}} = x$$

From properties of exponents, $2 \times \frac{1}{2} = 1$ is the actual exponent of *x*. Another example can be seen with $x^{\frac{4}{7}}$. The variable *x*, raised to four-sevenths, is equal to the seventh root of *x* to the fourth power: $\sqrt[7]{x^4}$. In general,

$$x^{\frac{1}{n}} = \sqrt[n]{x}$$

and

$$x^{\frac{m}{n}} = \sqrt[n]{x^m}$$

Negative exponents also involve fractions. Whereas y^3 can also be rewritten as $\frac{y^3}{1}$, y^{-3} can be rewritten as $\frac{1}{y^3}$. A negative exponent means the exponential expression must be moved to the opposite spot in a fraction to make the exponent positive. If the negative appears in the numerator, it moves to the denominator. If the negative appears in the denominator, it is moved to the numerator. In general, $a^{-n} = \frac{1}{a^n}$, and a^{-n} and a^n are reciprocals.

Take, for example, the following expression:

$$\frac{a^{-4}b^2}{c^{-5}}$$

Since *a* is raised to the negative fourth power, it can be moved to the denominator. Since *c* is raised to the negative fifth power, it can be moved to the numerator. The *b* variable is raised to the positive second power, so it does not move.

The simplified expression is as follows:

$$\frac{b^2c^5}{a^4}$$

In mathematical expressions containing exponents and other operations, the order of operations must be followed. *PEMDAS* states that exponents are calculated after any parenthesis AL, division, addition, and subtraction.

The Evaluation of Positive Rational Roots and Exponents

There are a few rules for working with exponents. For any numbers a, b, m, n, the following hold true:

$$a^1 = a$$

$$1^a = 1$$

$$a^0 = 1$$

$$a^m \times a^n = a^{m+n}$$

$$a^m \div a^n = a^{m-n}$$

$$(a^m)^n = a^{m \times n}$$

$$(a \times b)^m = a^m \times b^m$$

$$(a \div b)^m = a^m \div b^m$$

Any number, including a fraction, can be an exponent. The same rules apply.

Scientific Notation

Scientific Notation is used to represent numbers that are either very small or very large. For example, the distance to the sun is approximately 150,000,000,000 meters. Instead of writing this number with so many zeros, it can be written in scientific notation as 1.5×10^{11}meters. The same is true for very small numbers, but the exponent becomes negative. If the mass of a human cell is 0.000000000001 kilograms, that measurement can be easily represented by 1.0×10^{-12} kilograms. In both situations, scientific notation makes the measurement easier to read and understand. Each number is translated to an expression with one digit in the tens place times an expression corresponding to the zeros.

When two measurements are given and both involve scientific notation, it is important to know how these interact with each other:

- In addition and subtraction, the exponent on the ten must be the same before any operations are performed on the numbers. For example, $(1.3 \times 10^4) + (3.0 \times 10^3)$ cannot be added until one of the exponents on the ten is changed. The 3.0×10^3 can be changed to 0.3×10^4, then the 1.3 and 0.3 can be added. The answer comes out to be 1.6×10^4.

- For multiplication, the first numbers can be multiplied and then the exponents on the tens can be added. Once an answer is formed, it may have to be converted into scientific notation again depending on the change that occurred.
    - The following is an example of multiplication with scientific notation:

$$(4.5 \times 10^3) \times (3.0 \times 10^{-5}) = 13.5 \times 10^{-2}$$

    - Since this answer is not in scientific notation, the decimal is moved over to the left one unit, and 1 is added to the ten's exponent. This results in the final answer: 1.35×10^{-1}.
- For division, the first numbers are divided, and the exponents on the tens are subtracted. Again, the answer may need to be converted into scientific notation form, depending on the type of changes that occurred during the problem.
- *Order of magnitude* relates to scientific notation and is the total count of powers of 10 in a number. For example, there are 6 orders of magnitude in 1,000,000. If a number is raised by an order of magnitude, it is multiplied times 10. Order of magnitude can be helpful in estimating results using very large or small numbers. An answer should make sense in terms of its order of magnitude.
    - For example, if area is calculated using two dimensions with 6 orders of magnitude, because area involves multiplication, the answer should have around 12 orders of magnitude. Also, answers can be estimated by rounding to the largest place value in each number. For example, $5{,}493{,}302 \times 2{,}523{,}100$ can be estimated by $5 \times 3 = 15$ with 12 orders of magnitude.

Patterns, Relationships, and Algebra

Patterns and Sequences

Patterns

Patterns are an important part of mathematics. When mathematical calculations are completed repeatedly, patterns can be recognized. Recognizing patterns is an integral part of mathematics because it helps you understand relationships between different ideas. For example, a sequence of numbers can be given, and being able to recognize the relationship between the given numbers can help in completing the sequence.

For instance, given the sequence of numbers $7, 14, 21, 28, 35, \ldots$, the next number in the sequence would be 42. This is because the sequence lists all multiples of 7, starting at 7. Sequences can also be built from addition, subtraction, and division. Being able to recognize the relationship between the values that are given is the key to finding out the next number in the sequence.

Patterns within a sequence can come in 2 distinct forms. The items either repeat in a constant order, or the items change from one step to another in some consistent way. The core is the smallest unit, or number of items, that repeats in a repeating pattern. For example, the pattern oo▲oo▲o… has a core that is oo▲. Knowing only the core, the pattern can be extended. Knowing the number of steps in the core allows the identification of an item in each step without drawing/writing the entire pattern out. For example, suppose you must find the tenth item in the previous pattern. Because the core consists of three items (oo▲), the core repeats in multiples of 3. In other words, steps 3, 6, 9, 12, etc. will be ▲

completing the core with the core starting over on the next step. For the above example, the 9th step will be ▲ and the 10th will be o.

The most common patterns where each item changes from one step to the next are arithmetic and geometric sequences. In an arithmetic sequence, the items increase or decrease by a constant difference. In other words, the same thing is added or subtracted to each item or step to produce the next. To determine if a sequence is arithmetic, see what must be added or subtracted to step one to produce step two. Then, check if the same thing is added/subtracted to step two to produce step three. The same thing must be added/subtracted to step three to produce step four, and so on. Consider the pattern 13, 10, 7, 4, ... To get from step one (13) to step two (10) by adding or subtracting requires subtracting by 3. The next step is checking if subtracting 3 from step two (10) will produce step three (7), and subtracting 3 from step three (7) will produce step four (4). In this case, the pattern holds true. Therefore, this is an arithmetic sequence in which each step is produced by subtracting 3 from the previous step. To extend the sequence, 3 is subtracted from the last step to produce the next. The next three numbers in the sequence are 1, -2, -5.

A geometric sequence is one in which each step is produced by multiplying or dividing the previous step by the same number. To see if a sequence is geometric, decide what step one must be multiplied or divided by to produce step two. Then check if multiplying or dividing step two by the same number produces step three, and so on. Consider the pattern 2, 8, 32, 128, ... To get from step one (2) to step two (8) requires multiplication by 4. The next step determines if multiplying step two (8) by 4 produces step three (32), and multiplying step three (32) by 4 produces step four (128). In this case, the pattern holds true. Therefore, this is a geometric sequence in which each step is found by multiplying the previous step by 4. To extend the sequence, the last step is multiplied by 4 and repeated. The next three numbers in the sequence are 512; 2,048; 8,192.

Arithmetic and geometric sequences can also be represented by shapes. For example, an arithmetic sequence could consist of shapes with three sides, four sides, and five sides. A geometric sequence could consist of eight blocks, four blocks, and two blocks (each step is produced by dividing the number of blocks in the previous step by 2).

Relationships Between the Corresponding Terms of Two Numerical Patterns

When given two number patterns, the corresponding terms should be examined to determine if a relationship exists between them. Corresponding terms between patterns are the pairs of numbers which appear in the same step of the two sequences. Consider the following patterns 1, 2, 3, 4,... and 3, 6, 9, 12, ... The corresponding terms are: 1 and 3; 2 and 6; 3 and 9; and 4 and 12. To identify the relationship, each pair of corresponding terms is examined. You can also examine the possibilities of performing an operation (+, −, ×, ÷) to each sequence. In this case:

$$1 + 2 = 3 \text{ or } 1 \times 3 = 3$$

$$2 + 4 = 6 \text{ or } 2 \times 3 = 6$$

$$3 + 6 = 9 \text{ or } 3 \times 3 = 9$$

$$4 + 8 = 12 \text{ or } 4 \times 3 = 12$$

The pattern is that the number from the first sequence multiplied by 3 equals the number in the second sequence. By assigning each sequence a label (input and output) or variable (*x* and *y*), the relationship can be written as an equation. The first sequence represents the inputs, or *x*, and the second sequence represents the outputs, or *y*. So, the relationship can be expressed as: $y = 3x$.

Consider the following sets of numbers:

A	2	4	6	8
B	6	8	10	12

To write a rule for the relationship between the values for *a* and the values for *b*, the corresponding terms (2 and 6; 4 and 8; 6 and 10; 8 and 12) are examined. The possibilities for producing *b* from *a* are:

$2 + 4 = 6$ or $2 \times 3 = 6$

$4 + 4 = 8$ or $4 \times 2 = 8$

$6 + 4 = 10$

$8 + 4 = 12$ or $8 \times 1.5 = 12$

The pattern is that adding 4 to the value of *a* produces the value of *b*. The relationship can be written as the equation $a + 4 = b$.

Recognizing Equivalent Fractions and Mixed Numbers

The value of a fraction does not change if multiplying or dividing both the numerator and the denominator by the same number (other than 0). In other words, $\frac{x}{y} = \frac{a \times x}{a \times y} = \frac{x \div a}{y \div a}$, as long as *a* is not 0. This means that $\frac{2}{5} = \frac{4}{10}$, for example. If *x* and *y* are integers that have no common factors, then the fraction is said to be *simplified*. This means $\frac{2}{5}$ is simplified, but $\frac{4}{10}$ is not.

Often when working with fractions, the fractions need to be rewritten so that they all share a single denominator—this is called finding a *common denominator* for the fractions. Using two fractions, $\frac{a}{b}$ and $\frac{c}{d}$, the numerator and denominator of the left fraction can be multiplied by *d*, while the numerator and denominator of the right fraction can be multiplied by *b*. This provides the fractions $\frac{a \times d}{b \times d}$ and $\frac{c \times b}{d \times b}$ with the common denominator $b \times d$.

A fraction whose numerator is smaller than its denominator is called a *proper fraction*. A fraction whose numerator is bigger than its denominator is called an *improper fraction*. These numbers can be rewritten as a combination of integers and fractions, called a *mixed number*. For example, $\frac{6}{5} = \frac{5}{5} + \frac{1}{5} = 1 + \frac{1}{5}$, and can be written as $1\frac{1}{5}$.

Computation with Integers and Negative Rational Numbers

Integers are the whole numbers together with their negatives. They include numbers like 5, 24, 0, -6, and 15. They do not include fractions or numbers that have digits after the decimal point.

Rational numbers are all numbers that can be written as a fraction using integers. A *fraction* is written as $\frac{x}{y}$ and represents the quotient of *x* being divided by *y*. More practically, it means dividing the whole into *y* equal parts, then taking *x* of those parts.

Examples of rational numbers include $\frac{1}{2}$ and $\frac{5}{4}$. The number on the top is called the *numerator*, and the number on the bottom is called the *denominator*. Because every integer can be written as a fraction with a denominator of 1, (e.g. $\frac{3}{1} = 3$), every integer is also a rational number.

When adding integers and negative rational numbers, there are some basic rules to determine if the solution is negative or positive:

Adding two positive numbers results in a positive number: $3.3 + 4.8 = 8.1$

Adding two negative numbers results in a negative number: $(-8) + (-6) = -14$

Adding one positive and one negative number requires taking the absolute values and finding the difference between them. Then, the sign of the number that has the higher absolute value for the final solution is used.

For example, (-9) + 11, has a difference of absolute values of 2. The final solution is 2 because 11 has the higher absolute value. Another example is 9 + (-11), which has a difference of absolute values of 2. The final solution is -2 because 11 has the higher absolute value.

When subtracting integers and negative rational numbers, one has to change the problem to adding the opposite and then apply the rules of addition.

Subtracting two positive numbers is the same as adding one positive and one negative number.

For example, 4.9 – 7.1 is the same as 4.9 + (-7.1). The solution is -2.2 since the absolute value of -7.1 is greater. Another example is 8.5 – 6.4 which is the same as 8.5 + (-6.4). The solution is 2.1 since the absolute value of 8.5 is greater.

Subtracting a positive number from a negative number results in negative value.

For example, (-12) – 7 is the same as (-12) + (-7) with a solution of -19.

Subtracting a negative number from a positive number results in a positive value.

For example, 12 – (-7) is the same as 12 + 7 with a solution of 19.

For multiplication and division of integers and rational numbers, if both numbers are positive or both numbers are negative, the result is a positive value.

For example, (-1.7)(-4) has a solution of 6.8 since both numbers are negative values.

If one number is positive and another number is negative, the result is a negative value.

For example, (-15)/5 has a solution of -3 since there is one negative number.

Ratios and Proportional Relationships

A *ratio* compares the size of one group to the size of another. For example, there may be a room with 4 tables and 24 chairs. The ratio of tables to chairs is 4: 24. Such ratios behave like fractions in that both sides of the ratio by the same number can be multiplied or divided. Thus, the ratio 4:24 is the same as the ratio 2:12 and 1:6.

One quantity is *proportional* to another quantity if the first quantity is always some multiple of the second. For instance, the distance travelled in five hours is always five times to the speed as travelled. The distance is proportional to speed in this case.

One quantity is *inversely proportional* to another quantity if the first quantity is equal to some number divided by the second quantity. The time it takes to travel one hundred miles will be given by 100 divided by the speed travelled. The time is inversely proportional to the speed.

When dealing with word problems, there is no fixed series of steps to follow, but there are some general guidelines to use. It is important that the quantity to be found is identified. Then, it can be determined how the given values can be used and manipulated to find the final answer.

Ratios are used to show the relationship between two quantities. The ratio of oranges to apples in the grocery store may be 3 to 2. That means that for every 3 oranges, there are 2 apples. This comparison can be expanded to represent the actual number of oranges and apples. Another example may be the number of boys to girls in a math class. If the ratio of boys to girls is given as 2 to 5, that means there are 2 boys to every 5 girls in the class. Ratios can also be compared if the units in each ratio are the same. The ratio of boys to girls in the math class can be compared to the ratio of boys to girls in a science class by stating which ratio is higher and which is lower.

Rates are used to compare two quantities with different units. *Unit rates* are the simplest form of rate. With unit rates, the denominator in the comparison of two units is one. For example, if someone can type at a rate of 1000 words in 5 minutes, then his or her unit rate for typing is $\frac{1000}{5} = 200$ words in one minute or 200 words per minute. Any rate can be converted into a unit rate by dividing to make the denominator one. 1000 words in 5 minutes has been converted into the unit rate of 200 words per minute.

Ratios and rates can be used together to convert rates into different units. For example, if someone is driving 50 kilometers per hour, that rate can be converted into miles per hour by using a ratio known as the *conversion factor*. Since the given value contains kilometers and the final answer needs to be in miles, the ratio relating miles to kilometers needs to be used. There are 0.62 miles in 1 kilometer.

This, written as a ratio and in fraction form, is:

$$\frac{0.62\ miles}{1\ km}$$

To convert 50km/hour into miles per hour, the following conversion needs to be set up:

$$\frac{50\ km}{hour} * \frac{0.62\ miles}{1\ km} = 31\ miles\ per\ hour$$

The ratio between two similar geometric figures is called the *scale factor*. For example, a problem may depict two similar triangles, A and B. The scale factor from the smaller triangle A to the larger triangle B is given as 2 because the length of the corresponding side of the larger triangle, 16, is twice the corresponding side on the smaller triangle, 8. This scale factor can also be used to find the value of a missing side, x, in triangle A. Since the scale factor from the smaller triangle (A) to larger one (B) is 2, the larger corresponding side in triangle B (given as 25), can be divided by 2 to find the missing side in A (x = 12.5). The scale factor can also be represented in the equation $2A = B$ because two times the lengths of A gives the corresponding lengths of B. This is the idea behind similar triangles.

Much like a scale factor can be written using an equation like $2A = B$, a *relationship* is represented by the equation $Y = kX$. X and Y are proportional because as values of X increase, the values of Y also increase. A relationship that is inversely proportional can be represented by the equation $Y = \frac{k}{X'}$, where the value of Y decreases as the value of x increases and vice versa.

Proportional reasoning can be used to solve problems involving ratios, percentages, and averages. Ratios can be used in setting up proportions and solving them to find unknowns. For example, if a student completes an average of 10 pages of math homework in 3 nights, how long would it take the student to complete 22 pages? Both ratios can be written as fractions. The second ratio would contain the unknown.

The following proportion represents this problem, where x is the unknown number of nights:

$$\frac{10\ pages}{3\ nights} = \frac{22\ pages}{x\ nights}$$

Solving this proportion entails cross-multiplying and results in the following equation: $10x = 22 * 3$. Simplifying and solving for x results in the exact solution: $x = 6.6\ nights$. The result would be rounded up to 7 because the homework would actually be completed on the 7th night.

The following problem uses ratios involving percentages:

If 20% of the class is girls and 30 students are in the class, how many girls are in the class?

To set up this problem, it is helpful to use the common proportion:

$$\frac{\%}{100} = \frac{is}{of}$$

Within the proportion, % is the percentage of girls, 100 is the total percentage of the class, *is* is the number of girls, and *of* is the total number of students in the class. Most percentage problems can be written using this language. To solve this problem, the proportion should be set up as $\frac{20}{100} = \frac{x}{30'}$, and then solved for x. Cross-multiplying results in the equation $20 * 30 = 100x$, which results in the solution $x = 6$. There are 6 girls in the class.

Problems involving volume, length, and other units can also be solved using ratios. A problem may ask for the volume of a cone to be found that has a radius, $r = 7m$ and a height, $h = 16m$. Referring to the formulas provided on the test, the volume of a cone is given as:

$$V = \pi r^2 \frac{h}{3}$$

r is the radius, and h is the height. Plugging $r = 7$ and $h = 16$ into the formula, the following is obtained:

$$V = \pi(7^2)\frac{16}{3}$$

Therefore, volume of the cone is found to be approximately 821m^3. Sometimes, answers in different units are sought. If this problem wanted the answer in liters, 821m^3 would need to be converted.

Using the equivalence statement $1m^3$ = 1000L, the following ratio would be used to solve for liters:

$$821\text{m}^3 * \frac{1000L}{1m^3}$$

Cubic meters in the numerator and denominator cancel each other out, and the answer is converted to 821,000 liters, or $8.21 * 10^5$ L.

Other conversions can also be made between different given and final units. If the temperature in a pool is 30°C, what is the temperature of the pool in degrees Fahrenheit? To convert these units, an equation is used relating Celsius to Fahrenheit. The following equation is used:

$$T_{°F} = 1.8T_{°C} + 32$$

Plugging in the given temperature and solving the equation for T yields the result:

$$T_{°F} = 1.8(30) + 32 = 86°F$$

Both units in the metric system and U.S. customary system are widely used.

Example 1

Jana wants to travel to visit Alice, who lives one hundred and fifty miles away. If she can drive at fifty miles per hour, how long will her trip take?

The quantity to find is the *time* of the trip. The time of a trip is given by the distance to travel divided by the speed to be traveled. The problem determines that the distance is one hundred and fifty miles, while the speed is fifty miles per hour. Thus, 150 divided by 50 is $150 \div 50 = 3$. Because *miles* and *miles per hour* are the units being divided, the miles cancel out. The result is 3 hours.

Example 2

Bernard wishes to paint a wall that measures twenty feet wide by eight feet high. It costs ten cents to paint one square foot. How much money will Bernard need for paint?

The final quantity to compute is the *cost* to paint the wall. This will be ten cents ($0.10) for each square foot of area needed to paint. The area to be painted is unknown, but the dimensions of the wall are given; thus, it can be calculated.

The dimensions of the wall are 20 feet wide and 8 feet high. Since the area of a rectangle is length multiplied by width, the area of the wall is 8 x 20 = 160 square feet. Multiplying 0.1 x 160 yields $16 as the cost of the paint.

The *average* or *mean* of a collection of numbers is given by adding those numbers together and then dividing by the total number of values. A *weighted average* or *weighted mean* is given by adding the numbers multiplied by their weights, then dividing by the sum of the weights:

$$\frac{w_1x_1 + w_2x_2 + w_3x_3 \dots + w_nx_n}{w_1 + w_2 + w_3 + \cdots + w_n}$$

An *ordinary average* is a weighted average where all the weights are 1.

Algebra

Algebra is used to describe things in mathematics that have differing or changeable variables. It is easily applied to real world situations, due to its versatility. Algebra often uses variables, which represent unknown quantities or values. Variables are usually represented by a letter, such as X or Y. These are helpful when attempting to solve story problems. In algebra, letters are sometimes used to symbolize fixed values. In this case, the letters are called constants.

Below are some basic tips for navigating algebra.

To ensure multiplication signs (x) and unknown variables (X) are not confused, parentheses are placed around an object in an equation to signify multiplication.

Example
6 x 5 is the same as writing 6 (5).

Eliminate the multiplication sign between numbers and variables. It is understood they are multiplied.

Example
3 (X) and 3 x X and 3X all signify the same thing.

The multiplication symbol is sometimes replaced by a dot.

Example
6×5 can be written as $6 \cdot 5$.

Algebraic Expressions

Algebraic expressions look similar to equations, but they do not include the equal sign. Algebraic expressions are comprised of numbers, variables, and mathematical operations. Some examples of algebraic expressions are $8x + 7y - 12z$, $3a^2$, and $5x^3 - 4y^4$.

Algebraic expressions and equations can be used to represent real-life situations and model the behavior of different variables. For example, $2x + 5$ could represent the cost to play games at an arcade. In this case, 5 represents the price of admission to the arcade, and 2 represents the cost of each game played. To calculate the total cost, use the number of games played for x, multiply it by 2, and add 5.

Rewriting Expressions

Expressions can be rewritten based on their factors. For example, the expression $6x + 4$ can be rewritten as $2(3x + 2)$ because 2 is a factor of both $6x$ and 4. More complex expressions can also be rewritten based on their factors. The expression $x^4 - 16$ can be rewritten as $(x^2 - 4)(x^2 + 4)$. This is a different type of factoring, where a difference of squares is factored into a sum and difference of the same two terms. With some expressions, the factoring process is simple and only leads to a different way to represent the expression. With others, factoring and rewriting the expression leads to more information about the given problem.

In the following quadratic equation, factoring the binomial leads to finding the zeros of the function:

$$x^2 - 5x + 6 = y$$

This equations factors into $(x - 3)(x - 2) = y$, where 2 and 3 are found to be the zeros of the function when y is set equal to zero. The zeros of any function are the x-values where the graph of the function on the coordinate plane crosses the x-axis.

Factoring an equation is a simple way to rewrite the equation and find the zeros, but factoring is not possible for every quadratic. Completing the square is one way to find zeros when factoring is not an option. The following equation cannot be factored: $x^2 + 10x - 9 = 0$. The first step in this method is to move the constant to the right side of the equation, making it $x^2 + 10x = 9$. Then, the coefficient of x is divided by 2 and squared. This number is then added to both sides of the equation, to make the equation still true. For this example, $\left(\frac{10}{2}\right)^2 = 25$ is added to both sides of the equation to obtain:

$$x^2 + 10x + 25 = 9 + 25$$

This expression simplifies to $x^2 + 10x + 25 = 34$, which can then be factored into $(x + 5)^2 = 34$. Solving for x then involves taking the square root of both sides and subtracting 5. This leads to two zeros of the function:

$$x = \pm\sqrt{34} - 5$$

Depending on the type of answer the question seeks, a calculator may be used to find exact numbers.

Given a quadratic equation in standard form— $ax^2 + bx + c = 0$—the sign of a tells whether the function has a minimum value or a maximum value. If $a > 0$, the graph opens up and has a minimum value. If $a < 0$, the graph opens down and has a maximum value. Depending on the way the quadratic equation is written, multiplication may need to occur before a max/min value is determined.

Exponential expressions can also be rewritten, just as quadratic equations. Properties of exponents must be understood. Multiplying two exponential expressions with the same base involves adding the exponents:

$$a^m a^n = a^{m+n}$$

Dividing two exponential expressions with the same base involves subtracting the exponents:

$$\frac{a^m}{a^n} = a^{m-n}$$

Raising an exponential expression to another exponent includes multiplying the exponents:

$$(a^m)^n = a^{mn}$$

The zero power always gives a value of 1: $a^0 = 1$. Raising either a product or a fraction to a power involves distributing that power:

$$(ab)^m = a^m b^m \text{ and } \left(\frac{a}{b}\right)^m = \frac{a^m}{b^m}$$

Finally, raising a number to a negative exponent is equivalent to the reciprocal including the positive exponent:

$$a^{-m} = \frac{1}{a^m}$$

Linear Equations

A function is called *linear* if it can take the form of the equation $f(x) = ax + b$, or $y = ax + b$, for any two numbers *a* and *b*. A linear equation forms a straight line when graphed on the coordinate plane. An example of a linear function is shown below on the graph.

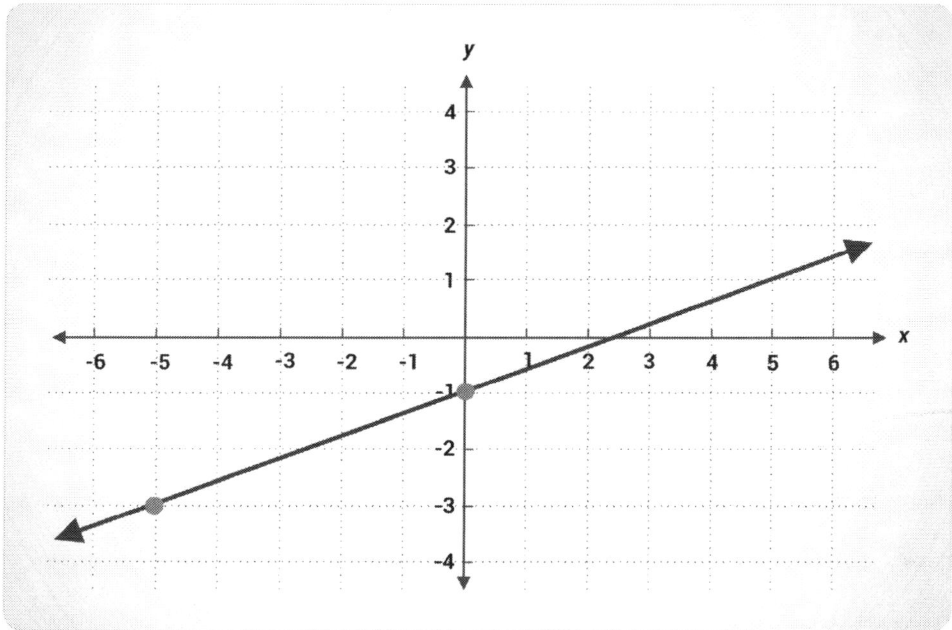

This is a graph of the following function: $y = \frac{2}{5}x - 1$. A table of values that satisfies this function is shown below.

x	y
-5	-3
0	-1
5	1
10	3

These points can be found on the graph using the form (x,y). For more on graphing in the coordinate plane, refer to the Graphing section on the next page.

Forms of Linear Equations

When graphing a linear function, note that the ratio of the change of the *y* coordinate to the change in the *x* coordinate is constant between any two points on the resulting line, no matter which two points are chosen. In other words, in a pair of points on a line, (x_1, y_1) and (x_2, y_2), with $x_1 \neq x_2$ so that the two points are distinct, then the ratio $\frac{y_2-y_1}{x_2-x_1}$ will be the same, regardless of which particular pair of points are chosen. This ratio, $\frac{y_2-y_1}{x_2-x_1}$, is called the *slope* of the line and is frequently denoted with the letter m. If slope *m* is positive, then the line goes upward when moving to the right, while if slope *m* is negative, then the line goes downward when moving to the right. If the slope is 0, then the line is called *horizontal*, and the *y* coordinate is constant along the entire line. In lines where the *x* coordinate is constant along the entire line, *y* is not actually a function of *x*. For such lines, the slope is not defined. These lines are called *vertical* lines.

Linear functions may take forms other than $y = ax + b$. The most common forms of linear equations are explained below:

- Standard Form: $\mathrm{Ax} + \mathrm{By} = \mathrm{C}$, in which the slope is given by $\mathrm{m} = \frac{-\mathrm{A}}{\mathrm{B}}$, and the y-intercept is given by $\frac{\mathrm{C}}{\mathrm{B}}$.
- Slope-Intercept Form: $\mathrm{y} = \mathrm{mx} + \mathrm{b}$, where the slope is m and the y intercept is b.
- Point-Slope Form: $\mathrm{y} - \mathrm{y}_1 = \mathrm{m}(\mathrm{x} - \mathrm{x}_1)$, where the slope is m and $(\mathrm{x}_1, \mathrm{y}_1)$ is any point on the chosen line.
- Two-Point Form: $\frac{\mathrm{y}-\mathrm{y}_1}{\mathrm{x}-\mathrm{x}_1} = \frac{\mathrm{y}_2-\mathrm{y}_1}{\mathrm{x}_2-\mathrm{x}_1}$, where $(\mathrm{x}_1, \mathrm{y}_1)$ and $(\mathrm{x}_2, \mathrm{y}_2)$ are any two distinct points on the chosen line. Note that the slope is given by $\mathrm{m} = \frac{\mathrm{y}_2-\mathrm{y}_1}{\mathrm{x}_2-\mathrm{x}_1}$.
- Intercept Form: $\frac{\mathrm{x}}{\mathrm{x}_1} + \frac{\mathrm{y}}{\mathrm{y}_1} = 1$, in which x_1 is the x-intercept and y_1 is the y-intercept.

These five ways to write linear equations are all useful in different circumstances. Depending on the given information, it may be easier to write one of the forms over another.

If $y = mx$, *y* is directly proportional to *x*. In this case, changing *x* by a factor changes *y* by that same factor. If $y = \frac{m}{x}$, *y* is inversely proportional to *x*. For example, if *x* is increased by a factor of 3, then *y* will be decreased by the same factor, 3.

Solving Linear Equations

Sometimes, rather than a situation where there's an equation such as $y = ax + b$ and finding *y* for some value of *x* is requested, the result is given and finding *x* is requested.

The key to solving any equation is to remember that from one true equation, another true equation can be found by adding, subtracting, multiplying, or dividing both sides by the same quantity. In this case, it's necessary to manipulate the equation so that one side only contains *x*. Then the other side will show what *x* is equal to.

For example, in solving $3x - 5 = 2$, adding 5 to each side results in $3x = 7$. Next, dividing both sides by 3 results in $x = \frac{7}{3}$. To ensure the value of x is correct, the number can be substituted into the original equation and solved to see if it makes a true statement. For example, $3(\frac{7}{3}) - 5 = 2$ can be simplified by cancelling out the two 3s. This yields $7 - 5 = 2$, which is a true statement.

Sometimes an equation may have more than one *x* term. For example, consider the following equation:

$$3x + 2 = x - 4$$

Moving all of the *x* terms to one side by subtracting *x* from both sides results in:

$$2x + 2 = -4$$

Next, subtract 2 from both sides so that there is no constant term on the left side. This yields $2x = -6$. Finally, divide both sides by 2, which leaves $x = -3$.

Graphing Functions and Relations

To graph relations and functions, the Cartesian plane is used. This means to think of the plane as being given a grid of squares, with one direction being the *x*-axis and the other direction the *y*-axis. Generally, the independent variable is placed along the horizontal axis, and the dependent variable is placed along the vertical axis. Any point on the plane can be specified by saying how far to go along the *x*-axis and how far along the *y*-axis with a pair of numbers (x, y). Specific values for these pairs can be given names such as $C = (-1, 3)$. Negative values mean to move left or down; positive values mean to move right or up. The point where the axes cross one another is called the *origin*. The origin has coordinates $(0, 0)$ and is usually called *O* when given a specific label.

An illustration of the Cartesian plane, along with the plotted points $(2, 1)$ and $(-1, -1)$, is below.

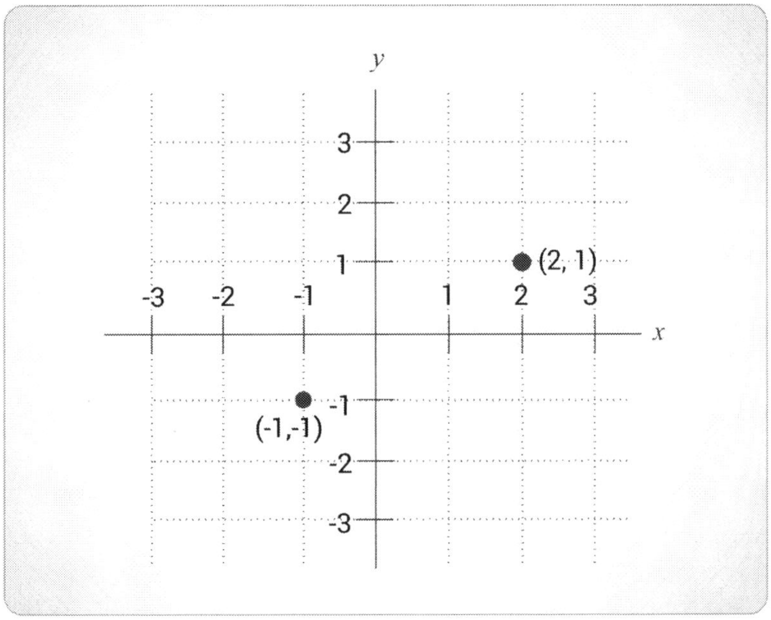

A *line plot* is a diagram that shows quantity of data along a number line. It is a quick way to record data in a structure similar to a bar graph without needing to do the required shading of a bar graph. Here is an example of a line plot:

A *tally chart* is a diagram in which tally marks are utilized to represent data. Tally marks are a means of showing a quantity of objects within a specific classification. Here is an example of a tally chart:

Number of days with rain	Number of weeks
0	II
1	𝍸
2	𝍸 I
3	𝍸 IIII
4	𝍸 𝍸 𝍸
5	𝍸
6	𝍸 I
7	IIII

Data is often recorded using fractions, such as half a mile, and understanding fractions is critical because of their popular use in real-world applications. Also, it is extremely important to label values with their units when using data. For example, regarding length, the number 2 is meaningless unless it is attached to a unit. Writing 2 cm shows that the number refers to the length of an object.

A *picture graph* is a diagram that shows pictorial representation of data being discussed. The symbols used can represent a certain number of objects. Notice how each fruit symbol in the following graph represents a count of two fruits. One drawback of picture graphs is that they can be less accurate if each symbol represents a large number. For example, if each banana symbol represented ten bananas, and students consumed 22 bananas, it may be challenging to draw and interpret two and one-fifth bananas as a frequency count of 22.

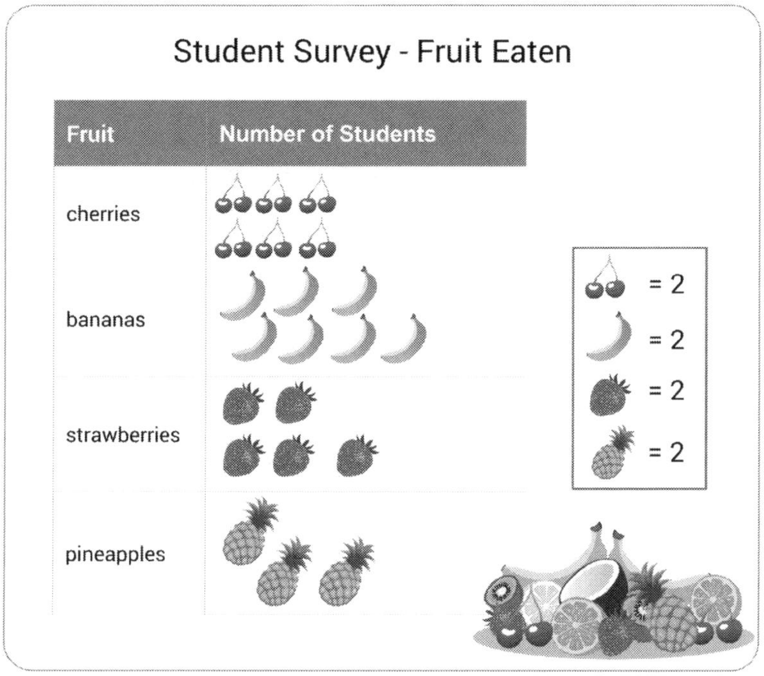

A circle graph, also called a pie chart, shows categorical data with each category representing a percentage of the whole data set. To make a circle graph, the percent of the data set for each category must be determined. To do so, the frequency of the category is divided by the total number of data points and converted to a percent. For example, if 80 people were asked what their favorite sport is and 20 responded basketball, basketball makes up 25% of the data ($\frac{20}{80} = 0.25 = 25\%$). Each category in a data set is represented by a *slice* of the circle proportionate to its percentage of the whole.

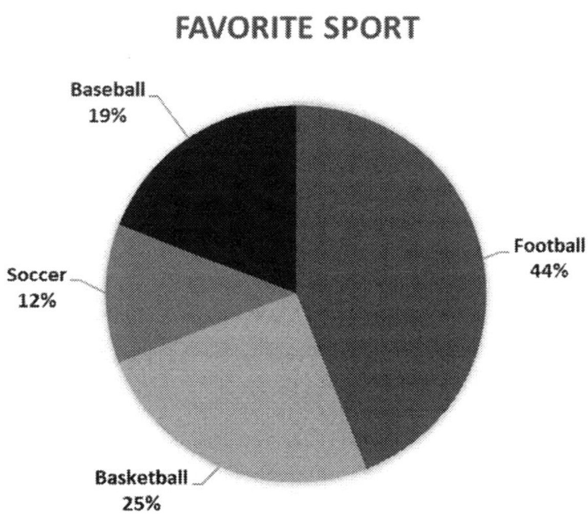

A scatter plot displays the relationship between two variables. Values for the independent variable, typically denoted by *x*, are paired with values for the dependent variable, typically denoted by *y*. Each set of corresponding values are written as an ordered pair (*x*, *y*). To construct the graph, a coordinate grid is labeled with the *x*-axis representing the independent variable and the *y*-axis representing the dependent variable. Each ordered pair is graphed.

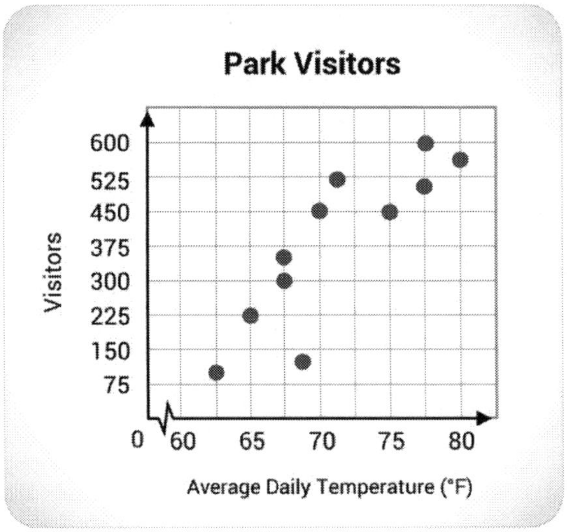

Like a scatter plot, a line graph compares two variables that change continuously, typically over time. Paired data values (ordered pair) are plotted on a coordinate grid with the *x*- and *y*-axis representing the two variables. A line is drawn from each point to the next, going from left to right. A double line graph simply displays two sets of data that contain values for the same two variables. The double line graph below displays the profit for given years (two variables) for Company A and Company B (two data sets).

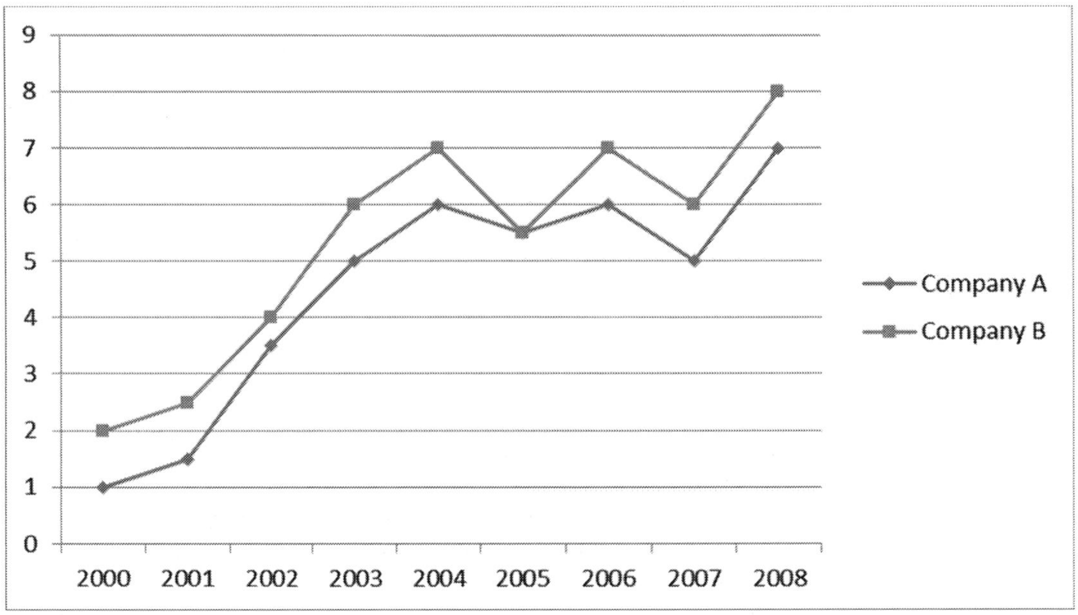

Choosing the appropriate graph to display a data set depends on what type of data is included in the set and what information must be shown.

Scatter plots and line graphs can be used to display data consisting of two variables. Examples include height and weight, or distance and time. A correlation between the variables is determined by examining the points on the graph. Line graphs are used if each value for one variable pairs with a distinct value for the other variable. Line graphs show relationships between variables.

Interpreting Competing Data

Be careful of questions with competing studies. These questions will ask the student to interpret which of two studies shows the greater amount or the higher rate of change between two results.

Here's an example. A research facility runs studies on two different antibiotics: Drug A and Drug B. The Drug A study includes 1,000 participants and cures 600 people. The Drug B study includes 200 participants and cures 150 people. Which drug is more successful?

The first step is to determine the percentage of each drug's rate of success. Drug A was successful in curing 60% of participants, while Drug B achieved a 75% success rate. Thus, Drug B is more successful based on these studies, even though it cured fewer people.

Sample size and experiment consistency should also be considered when answering questions based on competing studies. Is one study significantly larger than the other? In the antibiotics example, the Drug A study is five times larger than Drug B. Thus, Drug B's higher efficacy (desired result) could be a result of the smaller sample size, rather than the quality of drug.

Consistency between studies is directly related to sample size. Let's say the research facility elects to conduct more studies on Drug B. In the next study, there are 400 participants, and 200 are cured. The success rate of the second study is 50%. The results are clearly inconsistent with the first study, which means more testing is needed to determine the drug's efficacy. A hallmark of mathematical or scientific research is repeatability. Studies should be consistent and repeatable, with an appropriately large sample size, before drawing extensive conclusions.

What are Statistics?

The field of statistics describes relationships between quantities that are related, but not necessarily in a deterministic manner. For example, a graduating student's salary will often be higher when the student graduates with a higher GPA, but this is not always the case. Likewise, people who smoke tobacco are more likely to develop lung cancer, but, in fact, it is possible for non-smokers to develop the disease as well. *Statistics* describes these kinds of situations, where the likelihood of some outcome depends on the starting data.

Descriptive statistics involves analyzing a collection of data to describe its broad properties such average (or mean), what percent of the data falls within a given range, and other such properties. An example of this would be taking all of the test scores from a given class and calculating the average test score. *Inferential statistics* attempts to use data about a subset of some population to make inferences about the rest of the population. An example of this would be taking a collection of students who received tutoring and comparing their results to a collection of students who did not receive tutoring, then using that comparison to try to predict whether the tutoring program in question is beneficial.

To be sure that inferences have a high probability of being true for the whole population, the subset that is analyzed needs to resemble a miniature version of the population as closely as possible. For this reason, statisticians like to choose random samples from the population to study, rather than picking a specific group of people based on some similarity. For example, studying the incomes of people who live in Portland does not tell anything useful about the incomes of people who live in Tallahassee.

Mean, Median, and Mode

The center of a set of data (statistical values) can be represented by its mean, median, or mode. These are sometimes referred to as measures of central tendency.

Mean

The first property that can be defined for this set of data is the *mean*. This is the same as the average. To find the mean, add up all the data points, then divide by the total number of data points. For example, suppose that in a class of 10 students, the scores on a test were 50, 60, 65, 65, 75, 80, 85, 85, 90, 100. Therefore, the average test score will be:

$$\frac{50+60+65+65+75+80+85+85+90+100}{10} = 75.5$$

The mean is a useful number if the distribution of data is normal (more on this later), which roughly means that the frequency of different outcomes has a single peak and is roughly equally distributed on both sides of that peak. However, it is less useful in some cases where the data might be split or where there are some *outliers*. Outliers are data points that are far from the rest of the data. For example, suppose there are 10 executives and 90 employees at a company. The executives make $1000 per hour, and the employees make $10 per hour.

Therefore, the average pay rate will be:

$$\frac{\$1000 \times 10 + \$10 \times 90}{100} = \$109 \text{ per hour}$$

In this case, this average is not very descriptive since it's not close to the actual pay of the executives or the employees.

Median

Another useful measurement is the *median*. In a data set, the median is the point in the middle. The middle refers to the point where half the data comes before it and half comes after, when the data is recorded in numerical order. For instance, these are the speeds of the fastball of a pitcher during the last inning that he pitched (in order from least to greatest):

$$90, 92, 93, 93, 95, 96, 97, 97, 97$$

There are nine total numbers, so the middle or *median* number is the 5th one, which is 95.

In cases where the number of data points is an even number, then the average of the two middle points is taken. In the previous example of test scores, the two middle points are 75 and 80. Since there is no single point, the average of these two scores needs to be found. The average is:

$$\frac{75+80}{2} = 77.5$$

The median is generally a good value to use if there are a few outliers in the data. It prevents those outliers from affecting the "middle" value as much as when using the mean.

Since an outlier is a data point that is far from most of the other data points in a data set, this means an outlier also is any point that is far from the median of the data set. The outliers can have a substantial effect on the mean of a data set, but they usually do not change the median or mode, or do not change them by a large quantity. For example, consider the data set (3, 5, 6, 6, 6, 8). This has a median of 6 and a mode of 6, with a mean of $\frac{34}{6} \approx 5.67$. Now, suppose a new data point of 1000 is added so that the data set is now (3, 5, 6, 6, 6, 8, 1000). The median and mode, which are both still 6, remain unchanged. However, the average is now $\frac{1034}{7}$, which is approximately 147.7. In this case, the median and mode will be better descriptions for most of the data points.

The reason for outliers in a given data set is a complicated problem. It is sometimes the result of an error by the experimenter, but often they are perfectly valid data points that must be taken into consideration.

Mode

One additional measure to define for *X* is the *mode*. This is the data point that appears most frequently. If two or more data points all tie for the most frequent appearance, then each of them is considered a mode. In the case of the test scores, where the numbers were 50, 60, 65, 65, 75, 80, 85, 85, 90, 100, there are two modes: 65 and 85.

Describing a Set of Data

A set of data can be described in terms of its center, spread, shape and any unusual features. The center of a data set can be measured by its mean, median, or mode. The spread of a data set refers to how far the data points are from the center (mean or median). A data set with data points clustered around the center will have a small spread. A data set covering a wide range will have a large spread.

When a data set is displayed as a graph like the one below, the shape indicates if a sample is normally distributed, symmetrical, or has measures of skewness. When graphed, a data set with a normal distribution will resemble a bell curve.

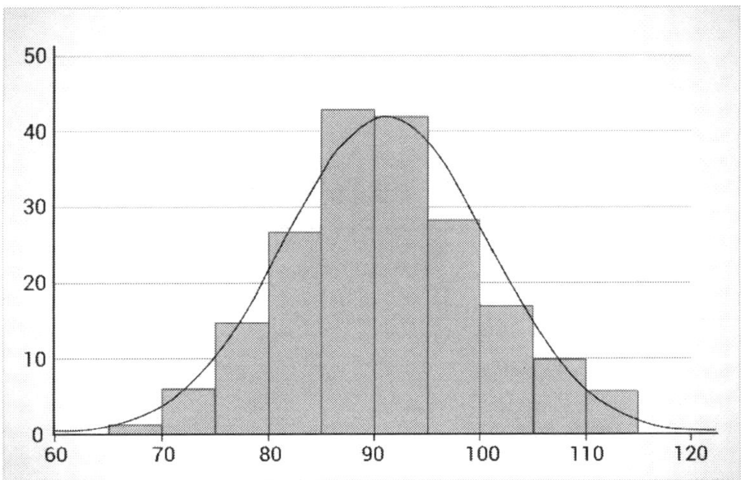

If the data set is symmetrical, each half of the graph when divided at the center is a mirror image of the other. If the graph has fewer data points to the right, the data is skewed right. If it has fewer data points to the left, the data is skewed left.

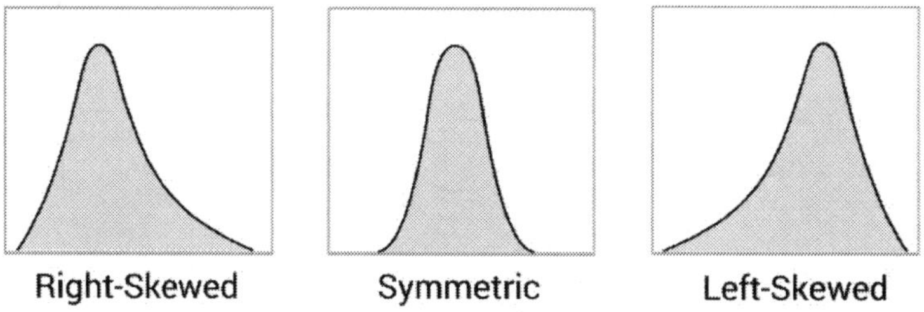

A description of a data set should include any unusual features such as gaps or outliers. A gap is a span within the range of the data set containing no data points. An outlier is a data point with a value either extremely large or extremely small when compared to the other values in the set.

The graphs above can be referred to as *unimodal* since they all have a single peak. In contrast, a bimodal graph has two peaks.

Correlation

An *X-Y diagram*, also known as a scatter diagram, visually displays the relationship between two variables. The independent variable is placed on the *x-axis*, or horizontal axis, and the dependent variable is placed on the *y-axis*, or vertical axis.

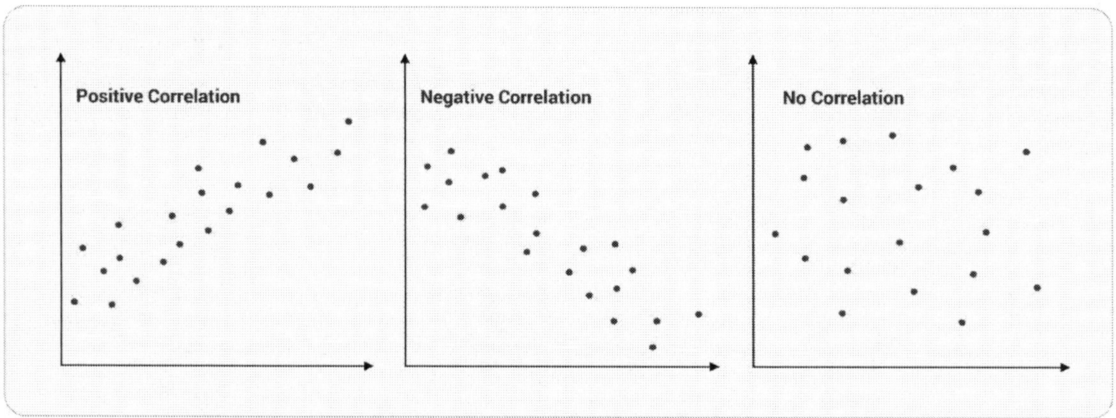

As shown in the figures above, an X-Y diagram may result in positive, negative, or no correlation between the two variables. In the first scatter plot, as the Y factor increases, the X factor increases as well. The opposite is true as well: as the X factor increases the Y factor also increases. Thus there is a positive correlation because one factor appears to positively affect the other factor

It's important to note, however, that a positive correlation between two variables doesn't equate to a cause-and-effect relationship. For example, a positive correlation between labor hours and units produced may not equate to a cause and effect relationship between the two. Any instance of correlation only indicates how likely the presence of one variable is in the instance of another. The variables should be further analyzed to determine which, if any, other variables (i.e. quality of employee work) may contribute to the positive correlation.

Comparing Data

Comparing data sets within statistics can mean many things. The first way to compare data sets is by looking at the center and spread of each set. The center of a data set can mean two things: median or mean. The *median* is the value that's halfway into each data set, and it splits the data into two intervals. The *mean* is the average value of the data within a set. It's calculated by adding up all of the data in the set and dividing the total by the number of data points. Outliers can significantly impact the mean. Additionally, two completely different data sets can have the same mean. For example, a data set with values ranging from 0 to 100 and a data set with values ranging from 44 to 56 can both have means of 50. The first data set has a much wider range, which is known as the *spread* of the data. This measures how varied the data is within each set.

Explaining the Relationship between Two Variables

In an experiment, variables are the key to analyzing data, especially when data is in a graph or table. Variables can represent anything, including objects, conditions, events, and amounts of time.

Covariance is a general term referring to how two variables move in relation to each other. Take for example an employee that gets paid by the hour. For them, hours worked and total pay have a positive covariance. As hours worked increases, so does pay.

Constant variables remain unchanged by the scientist across all trials. Because they are held constant for all groups in an experiment, they aren't being measured in the experiment, and they are usually ignored. Constants can either be controlled by the scientist directly like the nutrition, water, and sunlight given to plants, or they can be selected by the scientist specifically for an experiment like using a certain animal species or choosing to investigate only people of a certain age group.

Independent variables are also controlled by the scientist, but they are the same only for each group or trial in the experiment. Each group might be composed of students that all have the same color of car or each trial may be run on different soda brands. The independent variable of an experiment is what is being indirectly tested because it causes change in the dependent variables.

Dependent variables experience change caused by the independent variable and are what is being measured or observed. For example, college acceptance rates could be a dependent variable of an experiment that sorted a large sample of high school students by an independent variable such as test scores. In this experiment, the scientist groups the high school students by the independent variable (test scores) to see how it affects the dependent variable (their college acceptance rates).

Note that most variables can be held constant in one experiment, but also serve as the independent variable or a dependent variable in another. For example, when testing how well a fertilizer aids plant growth, its amount of sunlight should be held constant for each group of plants, but if the experiment is being done to determine the proper amount of sunlight a plant should have, the amount of sunlight is an independent variable because it is necessarily changed for each group of plants.

Probability

Given a set of possible outcomes *X*, a *probability distribution* on *X* is a function that assigns a probability to each possible outcome. If the outcomes are $(x_1, x_2, x_3, \dots x_n)$, and the probability distribution is *p*, then the following rules are applied.

- $0 \le p(x_i) \le 1$, for any i.
- $\sum_{i=1}^{n} p(x_i) = 1$.

In other words, the probability of a given outcome must be between zero and 1, while the total probability must be 1.

If $p(x_i)$ is constant, then this is called a *uniform probability distribution*, and $p(x_i) = \frac{1}{n}$. For example, on a six-sided die, the probability of each of the six outcomes will be $\frac{1}{6}$.

If seeking the probability of an outcome occurring in some specific range *A* of possible outcomes, written $P(A)$, add up the probabilities for each outcome in that range. For example, consider a six-sided die, and figure the probability of getting a 3 or lower when it is rolled. The possible rolls are 1, 2, 3, 4, 5, and 6. So, to get a 3 or lower, a roll of 1, 2, or 3 must be completed. The probabilities of each of these is $\frac{1}{6}$, so add these to get:

$$p(1) + p(2) + p(3) = \frac{1}{6} + \frac{1}{6} + \frac{1}{6} = \frac{1}{2}$$

Conditional Probabilities

An outcome occasionally lies within some range of possibilities *B*, and the probability that the outcomes also lie within some set of possibilities *A* needs to be figured. This is called a *conditional probability*. It is written as $P(A|B)$, which is read "the probability of *A* given *B*." The general formula for computing conditional probabilities is:

$$P(A|B) = \frac{P(A \cap B)}{P(B)}$$

However, when dealing with uniform probability distributions, simplify this a bit. Write $|A|$ to indicate the number of outcomes in *A*. Then, for uniform probability distributions, write:

$$P(A|B) = \frac{|A \cap B|}{|B|}$$

Recall that $A \cap B$ means "*A* intersect *B*," and consists of all of the outcomes that lie in both *A* and *B*.

This means that all possible outcomes do not need to be known. To see why this formula works, suppose that the set of outcomes *X* is $(x_1, x_2, x_3, \dots x_n)$, so that $|X| = n$. Then, for a uniform probability distribution:

$$P(A) = \frac{|A|}{n}$$

However, this means:

$$(A|B) = \frac{P(A \cap B)}{P(B)} = \frac{\frac{|A \cap B|}{n}}{\frac{|B|}{n}} = \frac{|A \cap B|}{|B|}$$

(since the *n*'s cancel out)

For example, suppose a die is rolled, and it is known that it will land between 1 and 4. However, how many sides the die has is unknown. Figure the probability that the die is rolled higher than 2. To figure this, $P(3)$ or $P(4)$ does not need to be determined, or any of the other probabilities, since it is known that a fair die has a uniform probability distribution. Therefore, apply the formula $\frac{|A \cap B|}{|B|}$. So, in this case *B* is (1, 2, 3, 4) and $A \cap B$ is (3, 4). Therefore:

$$\frac{|A \cap B|}{|B|} = \frac{2}{4} = \frac{1}{2}$$

Conditional probability is an important concept because, in many situations, the likelihood of one outcome can differ radically depending on how something else comes out. The probability of passing a test given that one has studied all of the material is generally much higher than the probability of passing a test given that one has not studied at all. The probability of a person having heart trouble is much lower if that person exercises regularly. The probability that a college student will graduate is higher when his or her SAT scores are higher, and so on. For this reason, there are many people who are interested in conditional probabilities.

Note that in some practical situations, changing the order of the conditional probabilities can make the outcome very different. For example, the probability that a person with heart trouble has exercised regularly is quite different than the probability that a person who exercises regularly will have heart trouble. The probability of a person receiving a military-only award, given that he or she is or was a soldier, is generally not very high, but the probability that a person being or having been a soldier, given that he or she received a military-only award, is 1.

However, in some cases, the outcomes do not influence one another this way. If the probability of *A* is the same regardless of whether *B* is given; that is, if $P(A|B) = P(A)$, then *A* and *B* are considered *independent*. In this case:

$$P(A|B) = \frac{P(A \cap B)}{P(B)} = P(A)$$

$$P(A \cap B) = P(A)P(B)$$

In fact, if $P(A \cap B) = P(A)P(B)$, it can be determined that $P(A|B) = P(A)$ and $P(A|B) = P(B)$ by working backward. Therefore, *B* is also independent of *A.*

An example of something being independent can be seen in rolling dice. In this case, consider a red die and a green die. It is expected that when the dice are rolled, the outcome of the green die should not depend in any way on the outcome of the red die. Or, to take another example, if the same die is rolled repeatedly, then the next number rolled should not depend on which numbers have been rolled previously. Similarly, if a coin is flipped, then the next flip's outcome does not depend on the outcomes of previous flips.

This can sometimes be counter-intuitive, since when rolling a die or flipping a coin, there can be a streak of surprising results. If, however, it is known that the die or coin is fair, then these results are just the result of the fact that over long periods of time, it is very likely that some unlikely streaks of outcomes will occur. Therefore, avoid making the mistake of thinking that when considering a series of independent outcomes, a particular outcome is "due to happen" simply because a surprising series of outcomes has already been seen.

There is a second type of common mistake that people tend to make when reasoning about statistical outcomes: the idea that when something of low probability happens, this is surprising. It would be surprising that something with low probability happened after just one attempt. However, with so much happening all at once, it is easy to see at least something happen in a way that seems to have a very low probability. In fact, a lottery is a good example. The odds of winning a lottery are very small, but the odds that somebody wins the lottery each week are actually fairly high. Therefore, no one should be surprised when some low probability things happen.

Addition Rule

The *addition rule* for probabilities states that the probability of *A* or *B* happening is:

$$P(A \cup B) = P(A) + P(B) - P(A \cap B)$$

Note that the subtraction of $P(A \cap B)$ must be performed, or else it would result in double counting any outcomes that lie in both *A* and in *B*. For example, suppose that a 20-sided die is being rolled. Fred bets that the outcome will be greater than 10, while Helen bets that it will be greater than 4 but less than 15. What is the probability that at least one of them is correct?

We apply the rule:

$$P(A \cup B) = P(A) + P(B) - P(A \cap B)$$

A is that outcome *x* is in the range $x > 10$, and *B* is that outcome *x* is in the range $4 < x < 15$.

$$P(A) = 10 \times \frac{1}{20} = \frac{1}{2}. P(B) = 10 \times \frac{1}{20} = \frac{1}{2}$$

$P(A \cap B)$ can be computed by noting that $A \cap B$ means the outcome *x* is in the range $10 < x < 15$, so:

$$P(A \cap B) = 4 \times \frac{1}{20} = \frac{1}{5}$$

Therefore:

$$P(A \cup B) = P(A) + P(B) - P(A \cap B) = \frac{1}{2} + \frac{1}{2} - \frac{1}{5} = \frac{4}{5}$$

Note that in this particular example, we could also have directly reasoned about the set of possible outcomes $A \cup B$, by noting that this would mean that *x* must be in the range $5 \leq x$. However, this is not always the case, depending on the given information.

Multiplication Rule

The *multiplication rule* for probabilities states the probability of *A* and *B* both happening is:

$$P(A \cap B) = P(A)P(B|A)$$

As an example, suppose that when Jamie wears black pants, there is a ½ probability that she wears a black shirt as well, and that she wears black pants ¾ of the time. What is the probability that she is wearing both a black shirt and black pants?

To figure this, use the above formula, where *A* will be "Jamie is wearing black pants," while *B* will be "Jamie is wearing a black shirt." It is known that $P(A)$ is ¾. It is also known that $P(B|A) = \frac{1}{2}$. Multiplying the two, the probability that she is wearing both black pants and a black shirt is:

$$P(A)P(B|A) = \frac{3}{4} \times \frac{1}{2} = \frac{3}{8}$$

Factorials

Factorials are a way of expressing the number of ways objects can be arranged. Factorial notation uses an exclamation mark after an expression to indicate that it is the product of that number and all positive integers less than it. For example, the expression 3! is equal to $3 \times 2 \times 1 = 6$, which corresponds with the number of ways 3 unique objects can be arranged or ordered. For example, if there are three blocks the letters A, B, and C, they can be arranged in the following six ways:

1	2	3	4	5	6
ABC	ACB	BAC	BCA	CAB	CBA

This is used frequently in statistics to find and express the number of ways multiple elements can be arranged. The general form for factorials is given as:

$$n! = n \times (n-1) \times (n-2) \ldots 1$$

Combinations and Permutations

There are many counting techniques that can help solve problems involving counting possibilities. For example, the *Addition Principle* states that if there are m choices from Group 1 and n choices from Group 2, then $n + m$ is the total number of choices possible from Groups 1 and 2. For this to be true, the groups can't have any choices in common. The *Multiplication Principle* states that if Process 1 can be completed n ways and Process 2 can be completed m ways, the total number of ways to complete both Process 1 and Process 2 is $n \times m$. For this rule to be used, both processes must be independent of each other. Counting techniques also involve permutations. A *permutation* is an arrangement of elements in a set for which order must be considered. For example, if three letters from the alphabet are chosen, ABC and BAC are two different permutations. The multiplication rule can be used to determine the total number of possibilities. If each letter can't be selected twice, the total number of possibilities is:

$$26 \times 25 \times 24 = 15{,}600$$

A formula can also be used to calculate this total. In general, the notation $P(n,r)$ represents the number of ways to arrange r objects from a set of n, and the formula is:

$$P(n,r) = \frac{n!}{(n-r)!}$$

In the previous example:

$$P(26,3) = \frac{26!}{23!} = 15{,}600$$

Contrasting permutations, a *combination* is an arrangement of elements in which order doesn't matter. In this case, ABC and BAC are the same combination. In the previous scenario, there are six permutations that represent each single combination. Therefore, the total number of possible combinations is:

$$15{,}600 \div 6 = 2{,}600$$

In general, $C(n,r)$ represents the total number of combinations of n items selected r at a time where order doesn't matter, and the formula is:

$$C(n,r) = \frac{n!}{(n-r)!\ r!}$$

Therefore, the following relationship exists between permutations and combinations:

$$C(n,r) = \frac{P(n,r)}{n!} = \frac{P(n,r)}{P(r,r)}$$

Geometry and Measurement

Plane Geometry

Locations on the plane that have no width or breadth are called *points*. These points usually will be denoted with capital letters such as *P*.

Any pair of points *A*, *B* on the plane will determine a unique straight line between them. This line is denoted *AB*. Sometimes to emphasize a line is being considered, this will be written as $\overleftrightarrow{AB}$.

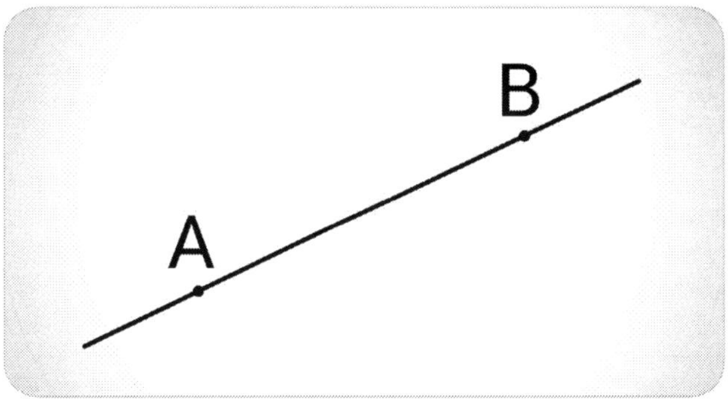

If the Cartesian coordinates for *A* and *B* are known, then the distance $d(A, B)$ along the line between them can be measured using the *Pythagorean formula*, which states that if:

$$A = (x_1, y_1) \text{ and } B = (x_2, y_2)$$

Then the distance between them is:

$$d(A, B) = \sqrt{(x_2 - x_1)^2 + (y_2 - y_1)^2}$$

The part of a line that lies between *A* and *B* is called a *line segment*. It has two endpoints, one at *A* and one at *B*. *Rays* also can be formed. Given points *A* and *B*, a *ray* is the portion of a line that starts at one of these points, passes through the other, and keeps on going. Therefore, a ray has a single endpoint, but the other end goes off to infinity.

Given a pair of points *A* and *B*, a circle centered at *A* and passing through *B* can be formed. This is the set of points whose distance from *A* is exactly $d(A, B)$. The radius of this circle will be $d(A, B)$.

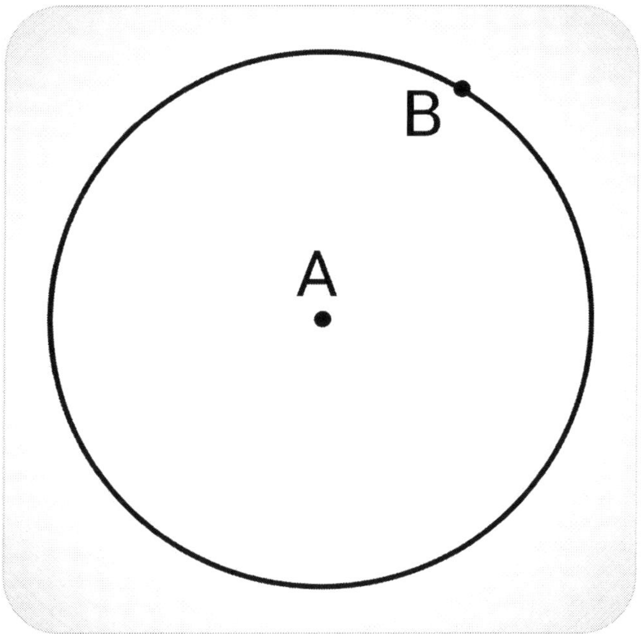

The *circumference* of a circle is the distance traveled by following the edge of the circle for one complete revolution, and the length of the circumference is given by $2\pi r$, where *r* is the radius of the circle. The formula for circumference is $C = 2\pi r$.

When two lines cross, they form an *angle*. The point where the lines cross is called the *vertex* of the angle. The angle can be named by either just using the vertex, $\angle A$, or else by listing three points $\angle BAC$, as shown in the diagram below.

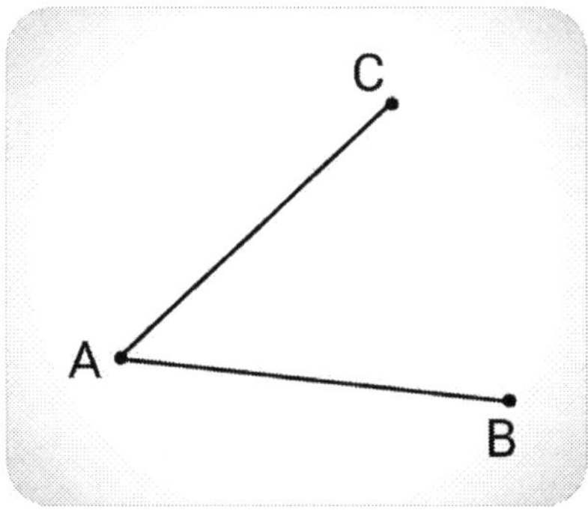

The measurement of an angle can be given in degrees or in radians. In degrees, a full circle is 360 degrees, written 360°. In radians, a full circle is 2π radians.

Given two points on the circumference of a circle, the path along the circle between those points is called an *arc* of the circle. For example, the arc between *B* and *C* is denoted by a thinner line:

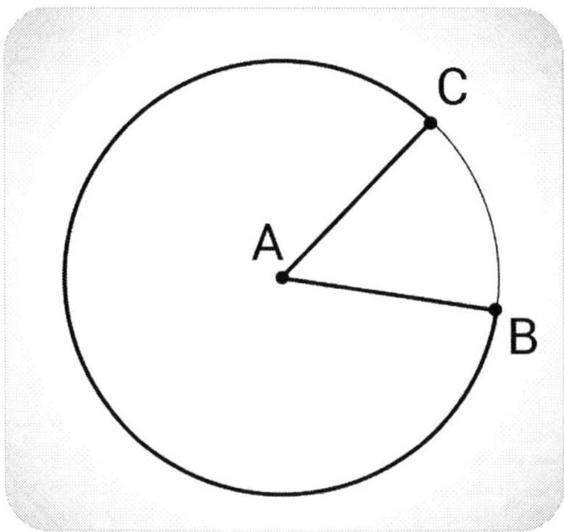

The length of the path along an arc is called the *arc length*. If the circle has radius *r*, then the arc length is given by multiplying the measure of the angle in radians by the radius of the circle.

Two lines are said to be *parallel* if they never intersect. If the lines are *AB* and *CD*, then this is written as $AB \parallel CD$.

If two lines cross to form four quarter-circles, that is, 90° angles, the two lines are *perpendicular*. If the point at which they cross is *B*, and the two lines are *AB* and *BC*, then this is written as $AB \perp BC$.

A *polygon* is a closed figure (meaning it divides the plane into an inside and an outside) consisting of a collection of line segments between points. These points are called the *vertices* of the polygon. These line segments must not overlap one another. Note that the number of sides is equal to the number of angles, or vertices of the polygon. The angles between line segments meeting one another in the polygon are called *interior angles*.

A *regular polygon* is a polygon whose edges are all the same length and whose interior angles are all of equal measure.

A *triangle* is a polygon with three sides. A *quadrilateral* is a polygon with four sides.

A *right triangle* is a triangle that has one 90° angle.

The sum of the interior angles of any triangle must add up to 180°.

An *isosceles triangle* is a triangle in which two of the sides are the same length. In this case, it will always have two congruent interior angles. If a triangle has two congruent interior angles, it will always be isosceles.

An *equilateral triangle* is a triangle whose sides are all the same length and whose angles are all equivalent to one another, equal to 60°. Equilateral triangles are examples of regular polygons. Note that equilateral triangles are also isosceles.

A *rectangle* is a quadrilateral whose interior angles are all 90°. A rectangle has two sets of sides that are equal to one another.

A *square* is a rectangle whose width and height are equal. Therefore, squares are regular polygons.

A *parallelogram* is a quadrilateral in which the opposite sides are parallel and equivalent to each other.

Transformations of a Plane

Given a figure drawn on a plane, many changes can be made to that figure, including *rotation*, *translation*, and *reflection*. Rotations turn the figure about a point, translations slide the figure, and reflections flip the figure over a specified line. When performing these transformations, the original figure is called the *pre-image*, and the figure after transformation is called the *image*.

More specifically, *translation* means that all points in the figure are moved in the same direction by the same distance. In other words, the figure is slid in some fixed direction. Of course, while the entire figure is slid by the same distance, this does not change any of the measurements of the figures involved. The result will have the same distances and angles as the original figure.

In terms of Cartesian coordinates, a translation means a shift of each of the original points (x, y) by a fixed amount in the *x* and *y* directions, to become $(x + a, y + b)$.

Another procedure that can be performed is called *reflection*. To do this, a line in the plane is specified, called the *line of reflection*. Then, take each point and flip it over the line so that it is the same distance from the line but on the opposite side of it. This does not change any of the distances or angles involved, but it does reverse the order in which everything appears.

To reflect something over the *x*-axis, the points (x, y) are sent to $(x, -y)$. To reflect something over the *y*-axis, the points (x, y) are sent to the points $(-x, y)$. Flipping over other lines is not something easy to express in Cartesian coordinates. However, by drawing the figure and the line of reflection, the distance to the line and the original points can be used to find the reflected figure.

Example: Reflect this triangle with vertices (-1, 0), (2, 1), and (2, 0) over the *y*-axis. The pre-image is shown below.

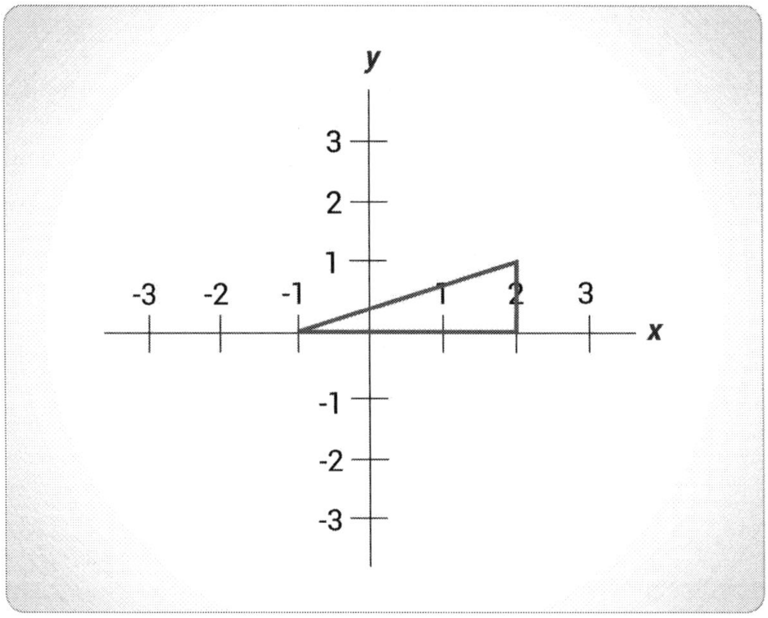

To do this, flip the *x* values of the points involved to the negatives of themselves, while keeping the *y* values the same. The image is shown here.

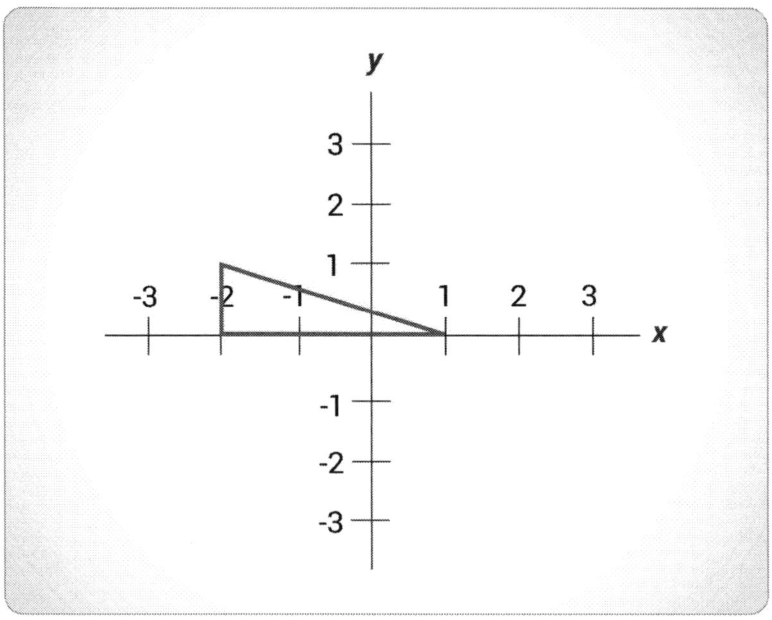

The new vertices will be (1, 0), (-2, 1), and (-2, 0).

Another procedure that does not change the distances and angles in a figure is *rotation*. In this procedure, pick a center point, then rotate every vertex along a circle around that point by the same angle. This procedure is also not easy to express in Cartesian coordinates, and this is not a requirement on this test. However, as with reflections, it's helpful to draw the figures and see what the result of the rotation would look like. This transformation can be performed using a compass and protractor.

Each one of these transformations can be performed on the coordinate plane without changes to the original dimensions or angles.

If two figures in the plane involve the same distances and angles, they are called *congruent figures*. In other words, two figures are congruent when they go from one form to another through reflection, rotation, and translation, or a combination of these.

Remember that rotation and translation will give back a new figure that is identical to the original figure, but reflection will give back a mirror image of it.

To recognize that a figure has undergone a rotation, check to see that the figure has not been changed into a mirror image, but that its orientation has changed (that is, whether the parts of the figure now form different angles with the *x* and *y* axes).

To recognize that a figure has undergone a translation, check to see that the figure has not been changed into a mirror image, and that the orientation remains the same.

To recognize that a figure has undergone a reflection, check to see that the new figure is a mirror image of the old figure.

Keep in mind that sometimes a combination of translations, reflections, and rotations may be performed on a figure.

Dilation

A *dilation* is a transformation that preserves angles, but not distances. This can be thought of as stretching or shrinking a figure. If a dilation makes figures larger, it is called an *enlargement*. If a dilation makes figures smaller, it is called a *reduction*. The easiest example is to dilate around the origin. In this case, multiply the *x* and *y* coordinates by a *scale factor*, k, sending points (x, y) to (kx, ky).

As an example, draw a dilation of the following triangle, whose vertices will be the points (-1, 0), (1, 0), and (1, 1).

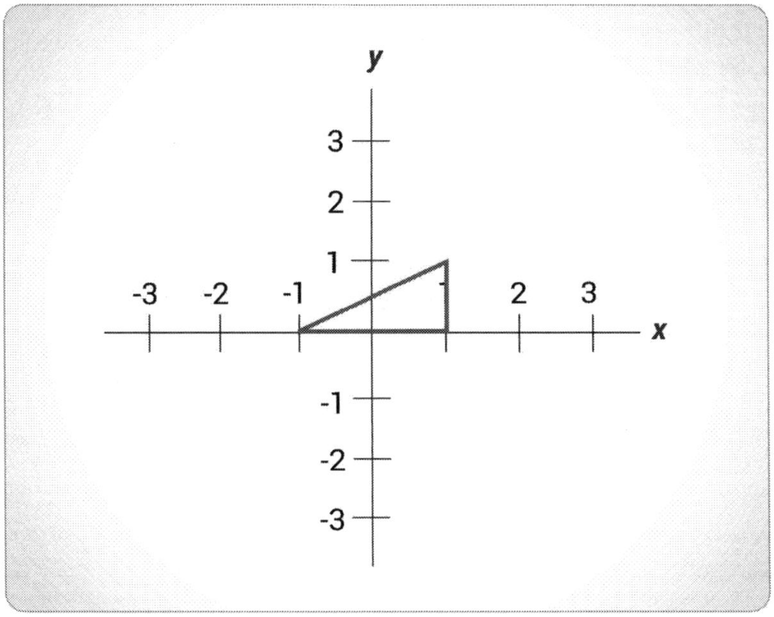

For this problem, dilate by a scale factor of 2, so the new vertices will be (-2, 0), (2, 0), and (2, 2).

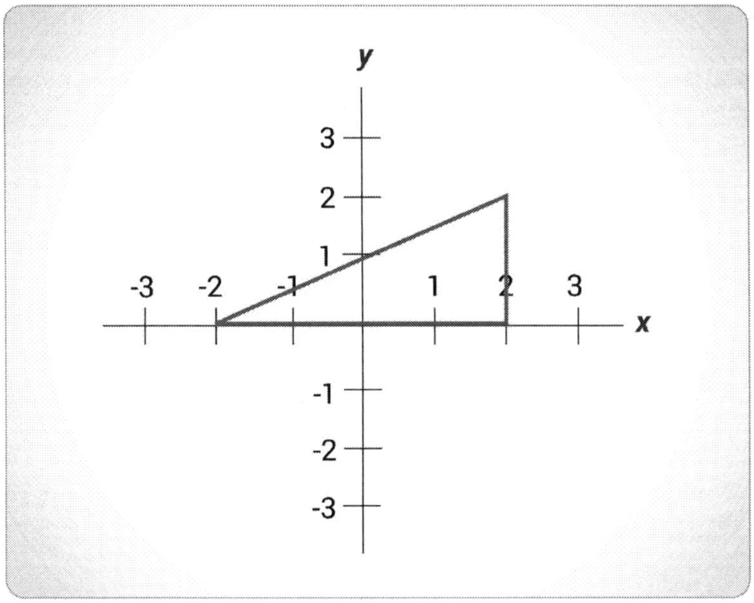

Note that after a dilation, the distances between the vertices of the figure will have changed, but the angles remain the same. The two figures that are obtained by dilation, along with possibly translation, rotation, and reflection, are all *similar* to one another. Another way to think of this is that similar figures have the same number of vertices and edges, and their angles are all the same. Similar figures have the same basic shape, but are different in size.

Symmetry

Using the types of transformations above, if an object can undergo these changes and not appear to have changed, then the figure is symmetrical. If an object can be split in half by a line and flipped over that line to lie directly on top of itself, it is said to have *line symmetry*. An example of both types of figures is seen below.

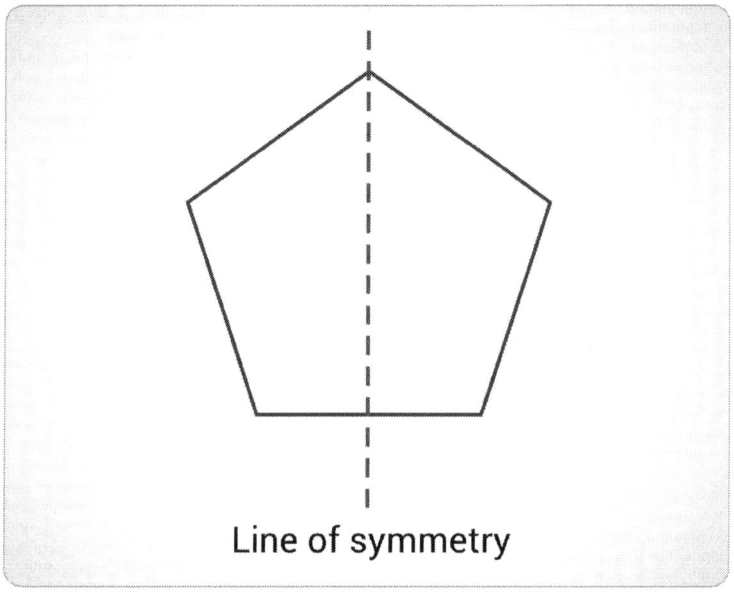

If an object can be rotated about its center to any degree smaller than 360, and it lies directly on top of itself, the object is said to have *rotational symmetry*. An example of this type of symmetry is shown below. The pentagon has an order of 5.

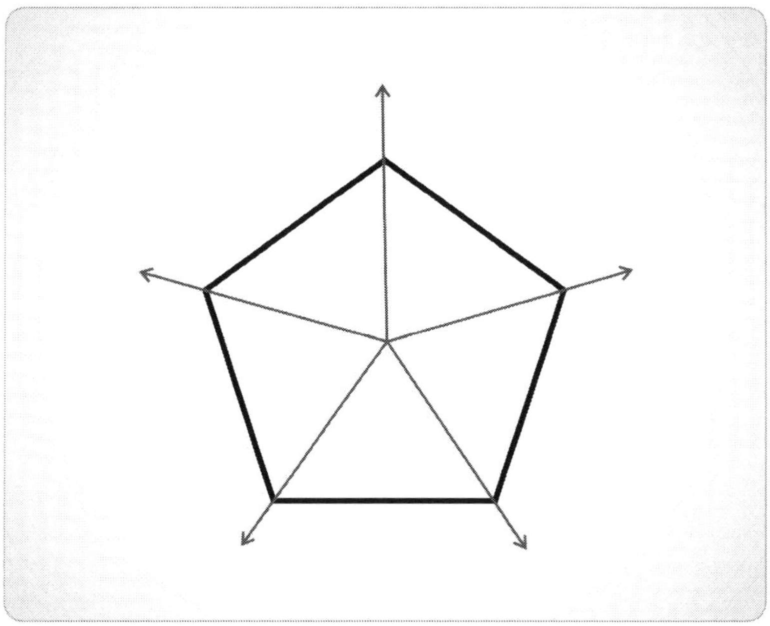

The rotational symmetry lines in the figure above can be used to find the angles formed at the center of the pentagon. Knowing that all of the angles together form a full circle, at 360 degrees, the figure can be split into 5 angles equally. By dividing the 360° by 5, each angle is 72°.

Given the length of one side of the figure, the perimeter of the pentagon can also be found using rotational symmetry. If one side length was 3 cm, that side length can be rotated onto each other side length four times. This would give a total of 5 side lengths equal to 3 cm. To find the perimeter, or distance around the figure, multiply 3 by 5. The perimeter of the figure would be 15 cm.

If a line cannot be drawn anywhere on the object to flip the figure onto itself or rotated less than or equal to 180 degrees to lay on top of itself, the object is asymmetrical. Examples of these types of figures are shown below.

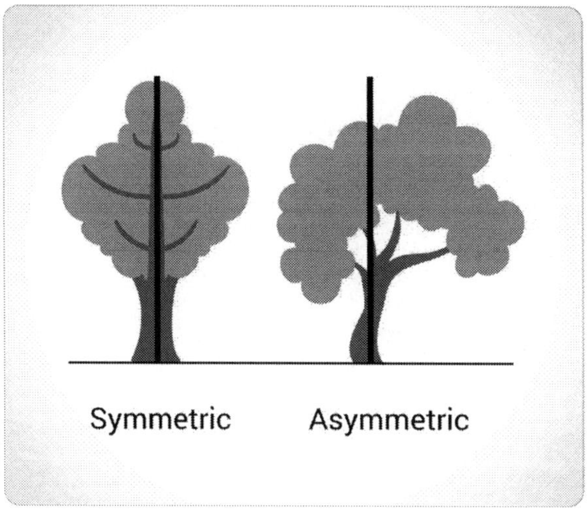

Perimeter and Area

Perimeter is the measurement of a distance around something or the sum of all sides of a polygon. Think of perimeter as the length of the boundary, like a fence. In contrast, *area* is the space occupied by a defined enclosure, like a field enclosed by a fence.

When thinking about perimeter, think about walking around the outside of something. When thinking about area, think about the amount of space or *surface area* something takes up.

Square

The perimeter of a square is measured by adding together all of the sides. Since a square has four equal sides, its perimeter can be calculated by multiplying the length of one side by 4. Thus, the formula is $P = 4 \times s$, where s equals one side. For example, the following square has side lengths of 5 meters:

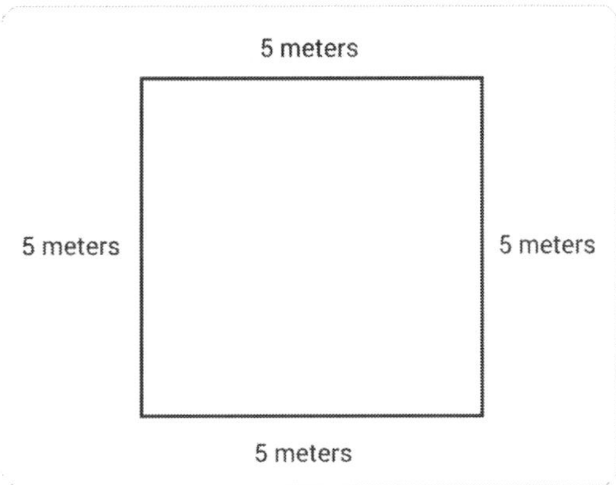

The perimeter is 20 meters because 4 times 5 is 20.

The area of a square is the length of a side squared. For example, if a side of a square is 7 centimeters, then the area is 49 square centimeters. The formula for this example is $A = s^2 = 7^2 = 49$ square centimeters. An example is if the rectangle has a length of 6 inches and a width of 7 inches, then the area is 42 square inches:

$$A = lw = 6(7) = 42 \text{ square inches}$$

Rectangle

Like a square, a rectangle's perimeter is measured by adding together all of the sides. But as the sides are unequal, the formula is different. A rectangle has equal values for its lengths (long sides) and equal values for its widths (short sides), so the perimeter formula for a rectangle is:

$$P = l + l + w + w = 2l + 2w$$

l equals length
w equals width

The area is found by multiplying the length by the width, so the formula is $A = l \times w$.

For example, if the length of a rectangle is 10 inches and the width 8 inches, then the perimeter is 36 inches because:

$$P = 2l + 2w = 2(10) + 2(8) = 20 + 16 = 36 \text{ inches}$$

Triangle

A triangle's perimeter is measured by adding together the three sides, so the formula is $P = a + b + c$, where a, b, and c are the values of the three sides. The area is the product of one-half the base and height so the formula is:

$$A = \frac{1}{2} \times b \times h$$

It can be simplified to:

$$A = \frac{bh}{2}$$

The base is the bottom of the triangle, and the height is the distance from the base to the peak. If a problem asks to calculate the area of a triangle, it will provide the base and height.

For example, if the base of the triangle is 2 feet and the height 4 feet, then the area is 4 square feet. The following equation shows the formula used to calculate the area of the triangle:

$$\mathrm{A} = \frac{1}{2}\mathrm{bh} = \frac{1}{2}(2)(4) = 4 \text{ square feet}$$

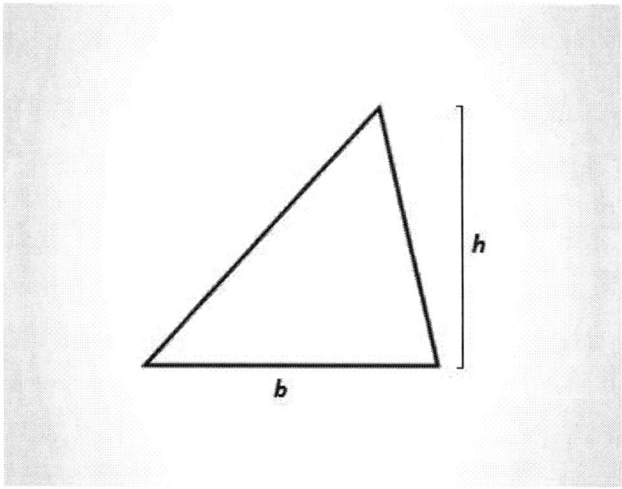

Circle

A circle's perimeter—also known as its circumference—is measured by multiplying the diameter by π.

Diameter is the straight line measured from a point on one side of the circle to a point directly across on the opposite side of the circle.

π is referred to as pi and is equal to 3.14 (with rounding).

So the formula is $\pi \times d$.

This is sometimes expressed by the formula $C = 2 \times \pi \times r$, where r is the radius of the circle. These formulas are equivalent, as the radius equals half of the diameter.

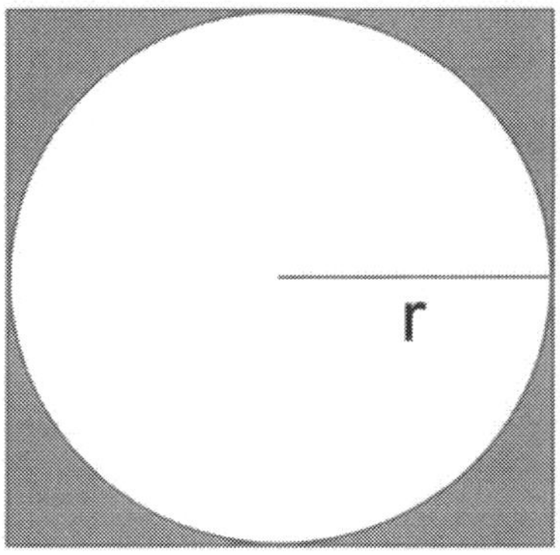

The area of a circle is calculated through the formula $A = \pi \times r^2$. The test will indicate either to leave the answer with π attached or to calculate to the nearest decimal place, which means multiplying by 3.14 for π.

To find the areas of more *general polygons*, it is usually easiest to break up the polygon into rectangles and triangles. For example, find the area of the following figure whose vertices are (-1, 0), (-1, 2), (1, 3), and (1, 0).

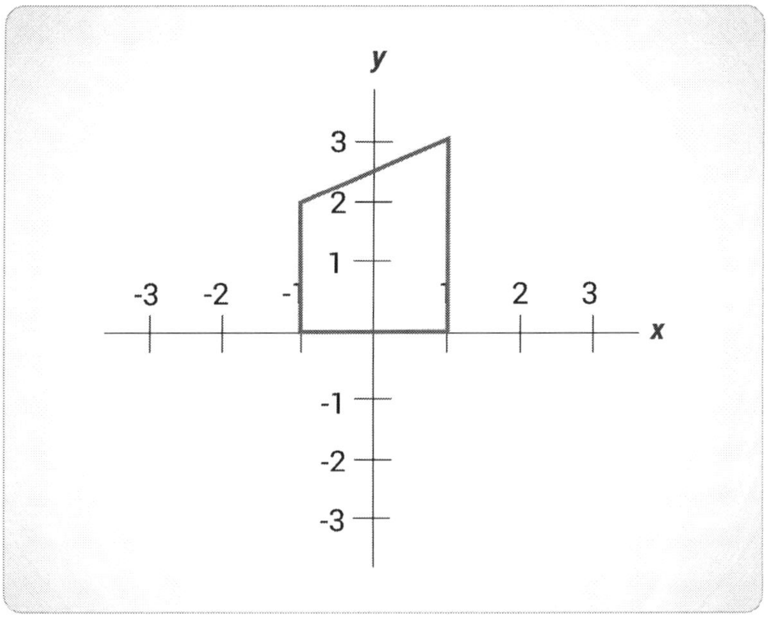

Separate this into a rectangle and a triangle as shown:

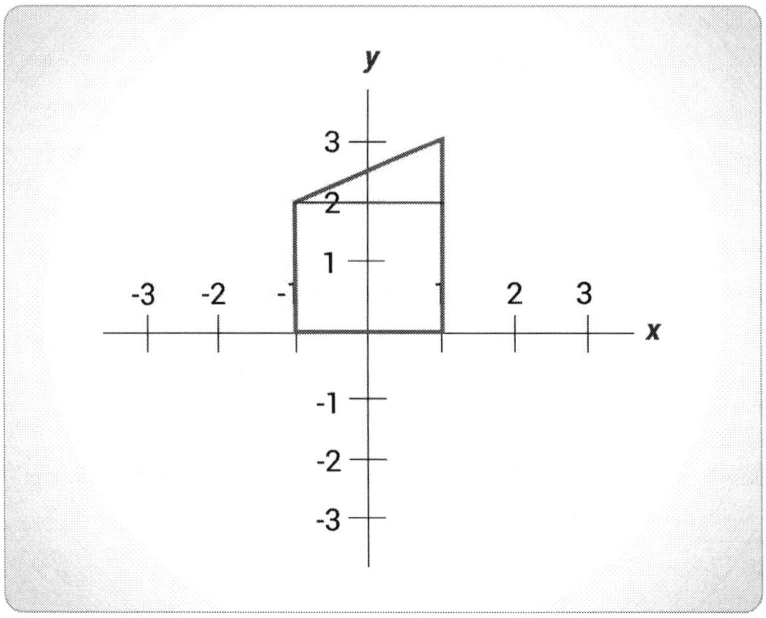

The rectangle has a height of 2 and a width of 2, so it has a total area of $2 \times 2 = 4$. The triangle has a width of 2 and a height of 1, so it has an area of $\frac{1}{2} 2 \times 1 = 1$. Therefore, the entire quadrilateral has an area of 4 + 1 = 5.

As another example, suppose someone wants to tile a rectangular room that is 10 feet by 6 feet using triangular tiles that are 12 inches by 6 inches. How many tiles would be needed? To figure this, first find the area of the room, which will be $10 \times 6 = 60$ square feet. The dimensions of the triangle are 1 foot by ½ foot, so the area of each triangle is:

$$\frac{1}{2} \times 1 \times \frac{1}{2} = \frac{1}{4} \text{ square feet}$$

Notice that the dimensions of the triangle had to be converted to the same units as the rectangle. Now, take the total area divided by the area of one tile to find the answer:

$$\frac{60}{\frac{1}{4}} = 60 \times 4 = 240 \text{ tiles required}$$

Irregular Shapes

The perimeter of an irregular polygon is found by adding the lengths of all of the sides. In cases where all of the sides are given, this will be very straightforward, as it will simply involve finding the sum of the provided lengths. Other times, a side length may be missing and must be determined before the perimeter can be calculated. Consider the example below:

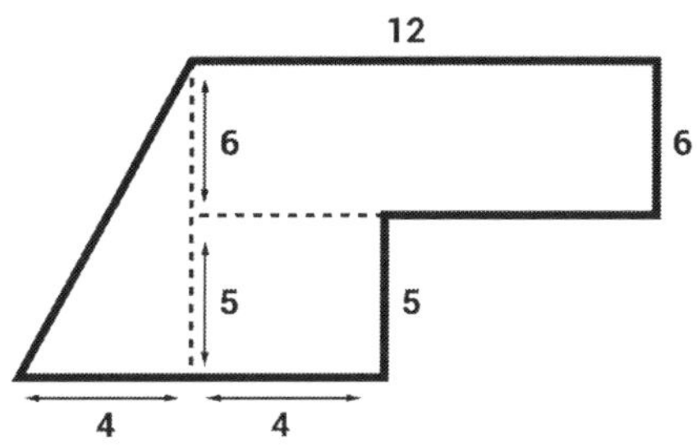

All of the side lengths are provided except for the angled side on the left. Test takers should notice that this is the hypotenuse of a right triangle. The other two sides of the triangle are provided (the base is 4 and the height is 6 + 5 = 11). The Pythagorean Theorem can be used to find the length of the hypotenuse, remembering that $a^2 + b^2 = c^2$.

Substituting the side values provided yields:

$$(4)^2 + (11)^2 = c^2$$

Therefore:

$$c = \sqrt{16 + 121} = 11.7$$

Finally, the perimeter can be found by adding this new side length with the other provided lengths to get the total length around the figure:

$$4 + 4 + 5 + 8 + 6 + 12 + 11.7 = 50.7$$

Although units are not provided in this figure, remember that reporting units with a measurement is important.

The area of an irregular polygon is found by decomposing, or breaking apart, the figure into smaller shapes. When the area of the smaller shapes is determined, these areas are added together to produce the total area of the area of the original figure. Consider the same example provided before:

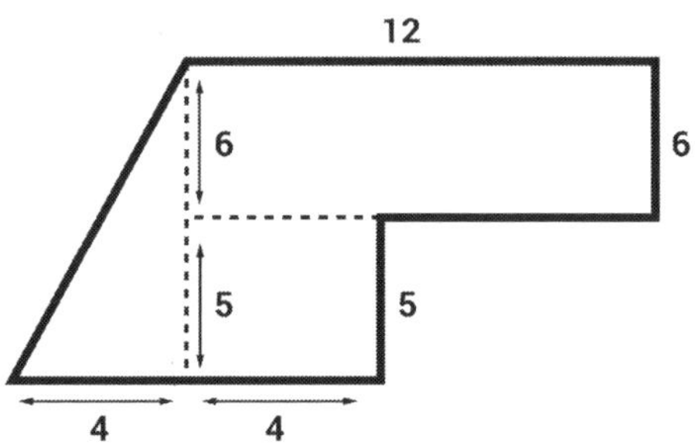

The irregular polygon is decomposed into two rectangles and a triangle. The area of the large rectangles ($A = l \times w \rightarrow A = 12 \times 6$) is 72 square units. The area of the small rectangle is 20 square units ($A = 4 \times 5$). The area of the triangle ($A = \frac{1}{2} \times b \times h \rightarrow A = \frac{1}{2} \times 4 \times 11$) is 22 square units. The sum of the areas of these figures produces the total area of the original polygon:

$$A = 72 + 20 + 22 \rightarrow A = 114 \text{ square units}$$

Arc

The *arc of a circle* is the distance between two points on the circle. The length of the arc of a circle in terms of *degrees* is easily determined if the value of the central angle is known. The length of the arc is simply the value of the central angle. In this example, the length of the arc of the circle in degrees is 75°.

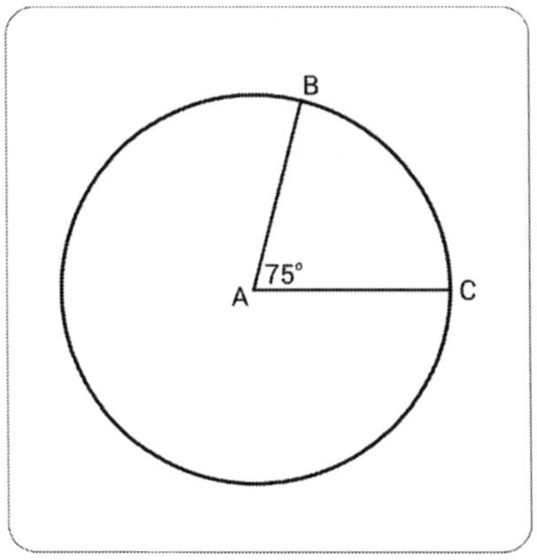

To determine the length of the arc of a circle in distance, the values for both the central angle and the radius must be known. This formula is:

$$\frac{central\ angle}{360°} = \frac{arc\ length}{2\pi r}$$

The equation is simplified by cross-multiplying to solve for the arc length.

In the following example, to solve for arc length, substitute the values of the central angle (75°) and the radius (10 inches) into the equation above.

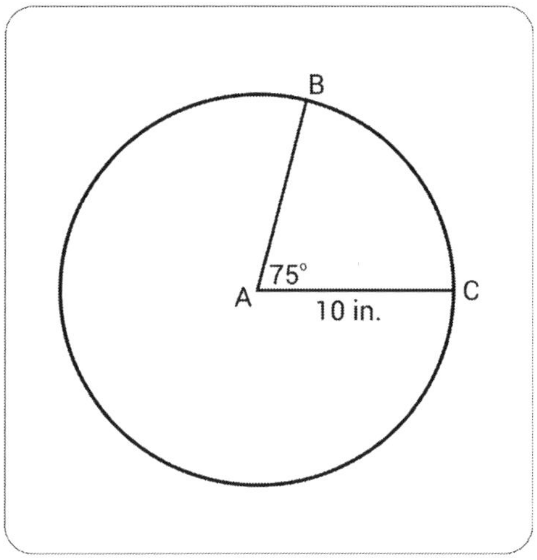

$$\frac{75°}{360°} = \frac{\text{arc length}}{2(3.14)(10\text{in.})}$$

To solve the equation, first cross-multiply: 4710 = 360(arc length). Next, divide each side of the equation by 360. The result of the formula is that the arc length is 13.1 (rounded).

Volumes and Surface Areas

Geometry in three dimensions is similar to geometry in two dimensions. The main new feature is that three points now define a unique *plane* that passes through each of them. Three dimensional objects can be made by putting together two-dimensional figures in different surfaces. Below, some of the possible three-dimensional figures will be provided, along with formulas for their volumes and surface areas.

A rectangular prism is a box whose sides are all rectangles meeting at 90° angles. Such a box has three dimensions: length, width, and height. If the length is *x*, the width is *y*, and the height is *z*, then the volume is given by $V = xyz$.

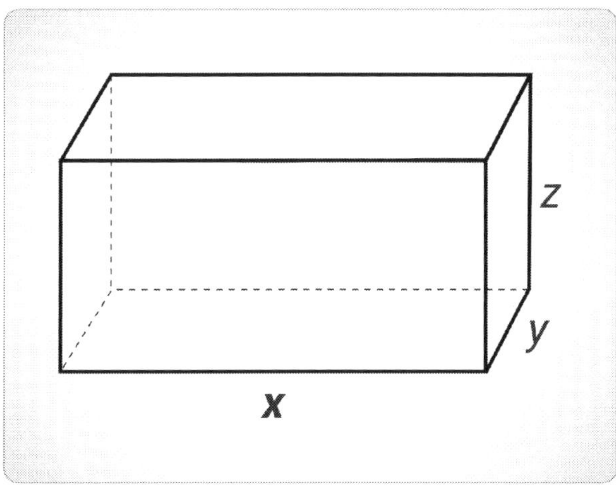

The surface area will be given by computing the surface area of each rectangle and adding them together. There are a total of six rectangles. Two of them have sides of length *x* and *y*, two have sides of length *y* and *z*, and two have sides of length *x* and *z*. Therefore, the total surface area will be given by:

$$SA = 2xy + 2yz + 2xz$$

A *rectangular pyramid* is a figure with a rectangular base and four triangular sides that meet at a single vertex. If the rectangle has sides of length *x* and *y*, then the volume will be given by:

$$V = \frac{1}{3}xyh$$

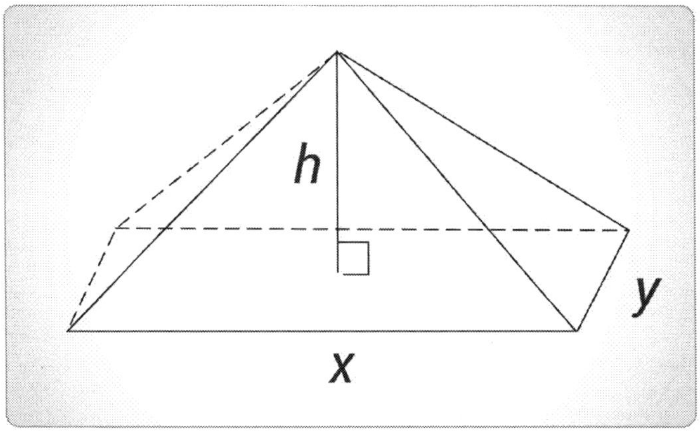

To find the surface area, the dimensions of each triangle need to be known. However, these dimensions can differ depending on the problem in question. Therefore, there is no general formula for calculating total surface area.

A *sphere* is a set of points all of which are equidistant from some central point. It is like a circle, but in three dimensions. The volume of a sphere of radius *r* is given by $V = \frac{4}{3}\pi r^3$. The surface area is given by $A = 4\pi r^2$.

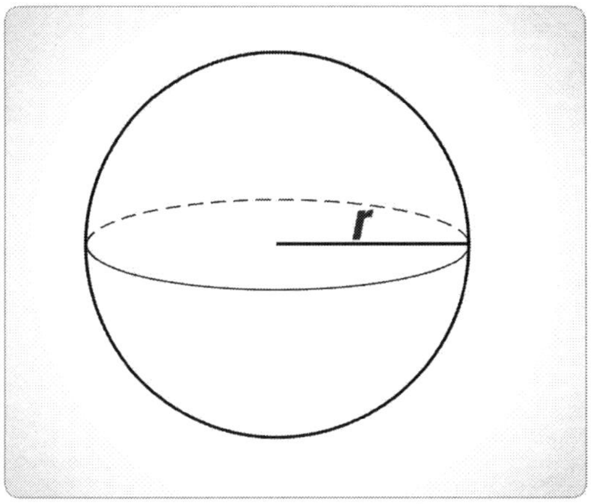

The Pythagorean Theorem

The Pythagorean theorem is an important tenet in geometry. It states that for right triangles, the sum of the squares of the two shorter sides will be equal to the square of the longest side (also called the *hypotenuse*). The longest side will always be the side opposite to the 90° angle. If this side is called *c*, and the other two sides are *a* and *b*, then the Pythagorean theorem states that $c^2 = a^2 + b^2$. Since lengths are always positive, this also can be written as:

$$c = \sqrt{a^2 + b^2}$$

A diagram to show the parts of a triangle using the Pythagorean theorem is below.

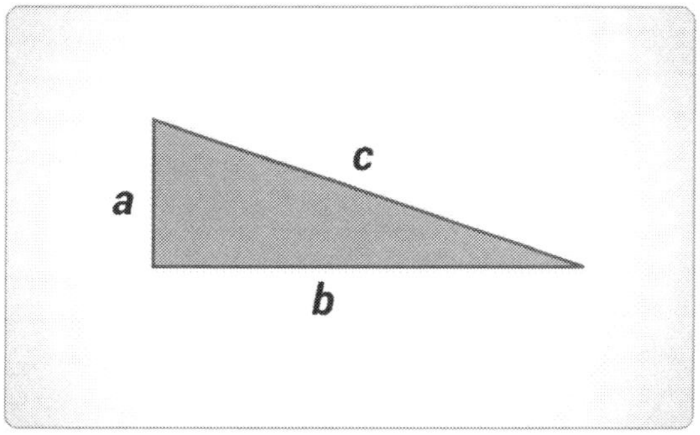

As an example of the theorem, suppose that Shirley has a rectangular field that is 5 feet wide and 12 feet long, and she wants to split it in half using a fence that goes from one corner to the opposite corner. How long will this fence need to be? To figure this out, note that this makes the field into two right triangles, whose hypotenuse will be the fence dividing it in half. Therefore, the fence length will be given by:

$$\sqrt{5^2 + 12^2} = \sqrt{169} = 13 \text{ feet long}$$

Similar Figures and Proportions

Sometimes, two figures are similar, meaning they have the same basic shape and the same interior angles, but they have different dimensions. If the ratio of two corresponding sides is known, then that ratio, or scale factor, holds true for all of the dimensions of the new figure.

Here is an example of applying this principle. Suppose that Lara is 5 feet tall and is standing 30 feet from the base of a light pole, and her shadow is 6 feet long. How high is the light on the pole? To figure this, it helps to make a sketch of the situation:

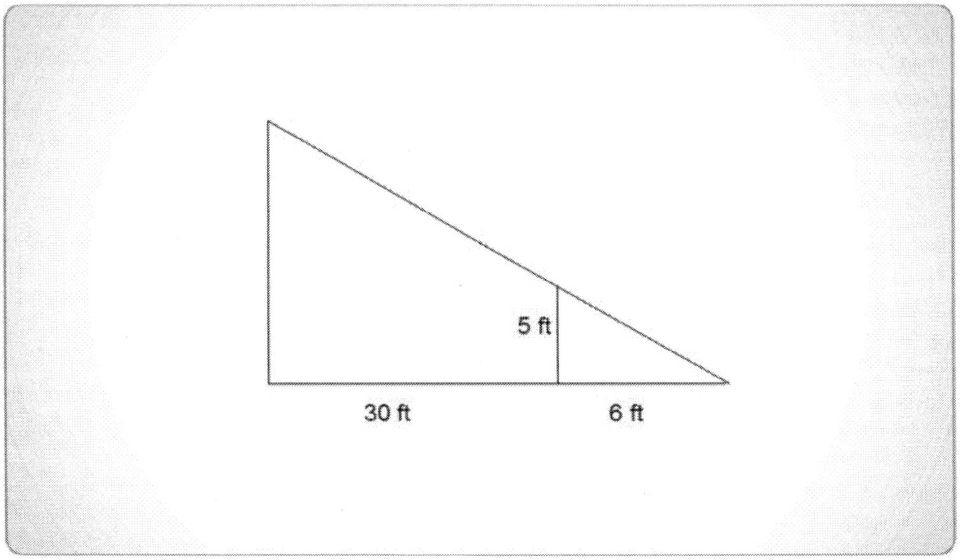

The light pole is the left side of the triangle. Lara is the 5-foot vertical line. Notice that there are two right triangles here, and that they have all the same angles as one another. Therefore, they form similar triangles. So, figure the ratio of proportionality between them.

The bases of these triangles are known. The small triangle, formed by Lara and her shadow, has a base of 6 feet. The large triangle, formed by the light pole along with the line from the base of the pole out to the end of Lara's shadow is $30 + 6 = 36$ feet long. So, the ratio of the big triangle to the little triangle will be $\frac{36}{6} = 6$. The height of the little triangle is 5 feet. Therefore, the height of the big triangle will be $6 \times 5 = 30$ feet, meaning that the light is 30 feet up the pole.

Notice that the perimeter of a figure changes by the ratio of proportionality between two similar figures, but the area changes by the *square* of the ratio. This is because if the length of one side is doubled, the area is quadrupled.

As an example, suppose two rectangles are similar, but the edges of the second rectangle are three times longer than the edges of the first rectangle. The area of the first rectangle is 10 square inches. How much more area does the second rectangle have than the first?

To answer this, note that the area of the second rectangle is $3^2 = 9$ times the area of the first rectangle, which is 10 square inches. Therefore, the area of the second rectangle is going to be $9 \times 10 = 90$ square inches. This means it has $90 - 10 = 80$ square inches more area than the first rectangle.

As a second example, suppose *X* and *Y* are similar right triangles. The hypotenuse of *X* is 4 inches. The area of *Y* is $\frac{1}{4}$ the area of *X*. What is the hypotenuse of *Y*?

First, realize the area has changed by a factor of $\frac{1}{4}$. The area changes by a factor that is the *square* of the ratio of changes in lengths, so the ratio of the lengths is the square root of the ratio of areas. That means that the ratio of lengths must be is $\sqrt{\frac{1}{4}} = \frac{1}{2}$, and the hypotenuse of *Y* must be $\frac{1}{2} \times 4 = 2$ inches.

Volumes between similar solids change like the cube of the change in the lengths of their edges. Likewise, if the ratio of the volumes between similar solids is known, the ratio between their lengths is known by finding the cube root of the ratio of their volumes.

For example, suppose there are two similar rectangular pyramids *X* and *Y*. The base of *X* is 1 inch by 2 inches, and the volume of *X* is 8 inches. The volume of *Y* is 64 inches. What are the dimensions of the base of *Y*?

To answer this, first find the ratio of the volume of *Y* to the volume of *X*. This will be given by $\frac{64}{8} = 8$. Now the ratio of lengths is the cube root of the ratio of volumes, or $\sqrt[3]{8} = 2$. So, the dimensions of the base of *Y* must be 2 inches by 4 inches.

Converting Within and Between Standard and Metric Systems

American Measuring System

The measuring system used today in the United States developed from the British units of measurement during colonial times. The most typically used units in this customary system are those used to measure weight, liquid volume, and length, whose common units are found below. In the customary system, the basic unit for measuring weight is the ounce (oz); there are 16 ounces (oz) in 1 pound (lb) and 2000 pounds in 1 ton. The basic unit for measuring liquid volume is the ounce (oz); 1 ounce is equal to 2 tablespoons (tbsp) or 6 teaspoons (tsp), and there are 8 ounces in 1 cup, 2 cups in 1 pint (pt), 2 pints in 1 quart (qt), and 4 quarts in 1 gallon (gal). For measurements of length, the inch (in) is the base unit; 12 inches make up 1 foot (ft), 3 feet make up 1 yard (yd), and 5280 feet make up 1 mile (mi). However, as there are only a set number of units in the customary system, with extremely large or extremely small amounts of material, the numbers can become awkward and difficult to compare.

Common Customary Measurements		
Length	**Weight**	**Capacity**
1 foot = 12 inches	1 pound = 16 ounces	1 cup = 8 fluid ounces
1 yard = 3 feet	1 ton = 2,000 pounds	1 pint = 2 cups
1 yard = 36 inches		1 quart = 2 pints
1 mile = 1,760 yards		1 quart = 4 cups
1 mile = 5,280 feet		1 gallon = 4 quarts
		1 gallon = 16 cups

Metric System

Aside from the United States, most countries in the world have adopted the metric system embodied in the International System of Units (SI). The three main SI base units used in the metric system are the meter (m), the kilogram (kg), and the liter (L); meters measure length, kilograms measure mass, and liters measure volume.

These three units can use different prefixes, which indicate larger or smaller versions of the unit by powers of ten. This can be thought of as making a new unit, which is sized by multiplying the original unit in size by a factor.

These prefixes and associated factors are:

Metric Prefixes			
Prefix	**Symbol**	**Multiplier**	**Exponential**
kilo	k	1,000	10^3
hecto	h	100	10^2
deca	da	10	10^1
no prefix		1	10^0
deci	d	0.1	10^{-1}
centi	c	0.01	10^{-2}
milli	m	0.001	10^{-3}

The correct prefix is then attached to the base. Some examples:

1 milliliter equals .001 liters.

1 kilogram equals 1,000 grams.

Choosing the Appropriate Measuring Unit

Some units of measure are represented as square or cubic units depending on the solution. For example, perimeter is measured in linear units, area is measured in square units, and volume is measured in cubic units.

Also be sure to use the most appropriate unit for the thing being measured. A building's height might be measured in feet or meters while the length of a nail might be measured in inches or centimeters. Additionally, for SI units, the prefix should be chosen to provide the most succinct available value. For example, the mass of a bag of fruit would likely be measured in kilograms rather than grams or milligrams, and the length of a bacteria cell would likely be measured in micrometers rather than centimeters or kilometers.

Conversion

Converting measurements in different units between the two systems can be difficult because they follow different rules. The best method is to look up an English to Metric system conversion factor and then use a series of equivalent fractions to set up an equation to convert the units of one of the measurements into those of the other. The table below lists some common conversion values that are useful for problems involving measurements with units in both systems:

English System	Metric System
1 inch	2.54 cm
1 foot	0.3048 m
1 yard	0.914 m
1 mile	1.609 km
1 ounce	28.35 g
1 pound	0.454 kg
1 fluid ounce	29.574 mL
1 quart	0.946 L
1 gallon	3.785 L

Consider the example where a scientist wants to convert 6.8 inches to centimeters. The table above is used to find that there are 2.54 centimeters in every inch, so the following equation should be set up and solved:

$$\frac{6.8\ in}{1} \times \frac{2.54\ cm}{1\ in} = 17.272\ cm$$

Notice how the inches in the numerator of the initial figure and the denominator of the conversion factor cancel out. (This equation could have been written simply as $6.8\ in \times 2.54\ cm = 17.272\ cm$, but it was shown in detail to illustrate the steps). The goal in any conversion equation is to set up the fractions so that the units you are trying to convert from cancel out and the units you desire remain.

For a more complicated example, consider converting 2.15 kilograms into ounces. The first step is to convert kilograms into grams and then grams into ounces. Note that the measurement you begin with does not have to be put in a fraction.

So in this case, 2.15 kg is by itself although it's technically the numerator of a fraction:

$$2.15\ kg \times \frac{1000g}{kg} = 2150\ g$$

Then, use the conversion factor from the table to convert grams to ounces:

$$2150g \times \frac{1\ oz}{28.35g} = 75.8\ oz$$

Practice Questions

1. Taylor was given several numbered pieces of paper for a party door prize. Any that get called could be redeemed for a choice of prizes, according to the value on the paper. Which of the following numbers has the greatest value?

   a. 1.4378
   b. 1.07548
   c. 1.43592
   d. 0.89409

2. The value of 6×12 is the same as:

   a. $2 \times 4 \times 4 \times 2$
   b. $7 \times 4 \times 3$
   c. $6 \times 6 \times 3$
   d. $3 \times 3 \times 4 \times 2$

3. This chart indicates how many sales of CDs, vinyl records, and MP3 downloads occurred over the last year. Approximately what percentage of the total sales was from CDs?

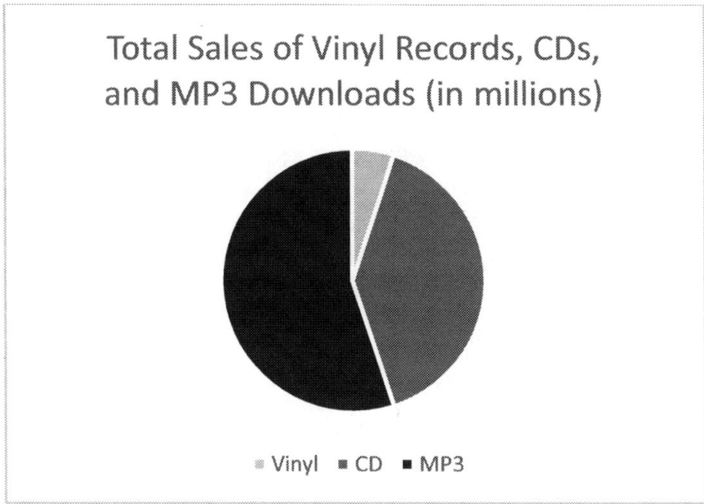

   a. 55%
   b. 25%
   c. 40%
   d. 5%

4. Frank has been saving up money in order to replace the old appliances in his house. After a 20% sale discount, Frank purchased a new refrigerator for $850. How much did he save from the original price?

   a. $170
   b. $212.50
   c. $105.75
   d. $200

5. What is the simplified form of the expression $1.2 \times 10^{12} \div 3.0 \times 10^{8}$?
   a. 0.4×10^{4}
   b. 4.0×10^{4}
   c. 4.0×10^{3}
   d. 3.6×10^{20}

6. For the expression $5!$, which of the following correctly evaluates the expression?
   a. 125
   b. 15
   c. 120
   d. 25

7. What is the value of x for the right triangle shown below?

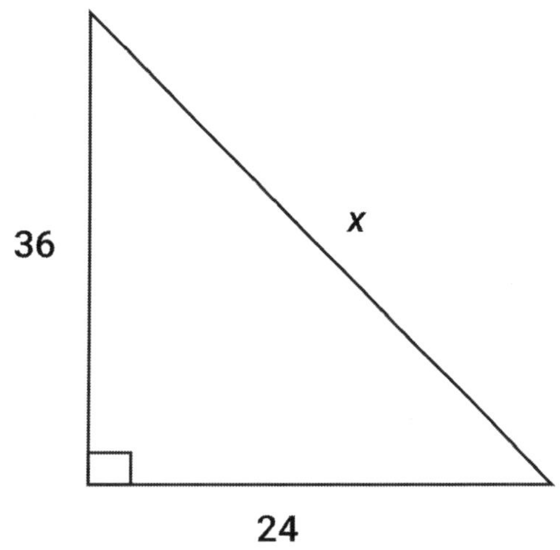

   a. 43.3
   b. 26.8
   c. 42.7
   d. 44.1

8. Express the solution to the following problem in decimal form:

$$\frac{3}{5} \times \frac{7}{10} \div \frac{1}{2}$$

   a. 0.042
   b. 84%
   c. 0.84
   d. 0.42

9. A toddler is playing with wooden shapes on the floor and starts making a pattern out of them. If the pattern was expanded, what would be the 42nd item in the following pattern: ▲○○□▲○○□▲…?

a. ○

b. ▲

c. □

d. None of the above

10. If Sarah reads at an average rate of 21 pages in four nights, how long will it take her to read 140 pages?

a. 6 nights

b. 26 nights

c. 8 nights

d. 27 nights

11. Alan currently weighs 200 pounds, but he wants to lose weight to get down to 175 pounds. What is this difference in kilograms? (1 pound is approximately equal to 0.45 kilograms.)

a. 9 kg

b. 11.25 kg

c. 78.75 kg

d. 90 kg

12. Johnny earns $2334.50 from his job each month. He pays $1437 for monthly expenses. Johnny is planning a vacation in 3 months' time that he estimates will cost $1750 total. How much will Johnny have left over from three months' of saving once he pays for his vacation?

a. $948.50

b. $584.50

c. $852.50

d. $942.50

13. Sarah has collected 420 miniature figurines from a vendor she passes every day on the way home. There are 98 figurines in a set. Sarah wants to know, approximately, how many full sets she could possibly have. What is $\frac{420}{98}$ rounded to the nearest integer?

a. 3

b. 4

c. 5

d. 6

14. What value of x makes the following equality true?

$$\sqrt{3x+5} - 2 = 1$$

a. $\frac{3}{5}$

b. $\frac{4}{3}$

c. 5

d. $\frac{1}{2}$

15. A house-shaped decoration used as a bookend has a rectangular base. The total perimeter of the base is 36 cm. If the length is 12 cm, what is the width of each side?

a. 3 cm
b. 12 cm
c. 6 cm
d. 8 cm

16. Dwayne has received the following scores on his math tests: 78, 92, 83, 97. What score must Dwayne get on his next math test to have an overall average of 90?

a. 89
b. 98
c. 95
d. 100

17. What is the overall median of Dwayne's current scores: 78, 92, 83, 97?

a. 19
b. 85
c. 83
d. 87.5

18. If three squared is subtracted from the product of the square root of thirty-six and the square root of sixteen, what is its value?

a. 30
b. 21
c. 15
d. 13

19. What is the probability of randomly picking the winner and runner-up from a race of 4 horses and distinguishing which is the winner?

a. $\frac{1}{4}$
b. $\frac{1}{2}$
c. $\frac{1}{16}$
d. $\frac{1}{12}$

20. If the square of the difference between twenty-five and twenty-one is divided by two and added to four times seven, what is the value of that expression?

a. 512
b. 36
c. 60.5
d. 22

21. Kimberley earns $10 an hour babysitting, and after 10 p.m., she earns $12 an hour, with the amount paid being rounded to the nearest hour accordingly. On her last job, she worked from 5:30 p.m. to 11 p.m. In total, how much did Kimberley earn on her last job?
   a. $45
   b. $57
   c. $62
   d. $42

22. What value of x would solve the following equation?

$$9x + x - 7 = 16 + 2x$$

   a. $x = -4$
   b. $x = 3$
   c. $x = \frac{9}{8}$
   d. $x = \frac{23}{8}$

23. Arrange the following numbers from least to greatest value:
   $0.85, \frac{4}{5}, \frac{2}{3}, \frac{91}{100}$
   a. $0.85, \frac{4}{5}, \frac{2}{3}, \frac{91}{100}$
   b. $\frac{4}{5}, 0.85, \frac{91}{100}, \frac{2}{3}$
   c. $\frac{2}{3}, \frac{4}{5}, 0.85, \frac{91}{100}$
   d. $0.85, \frac{91}{100}, \frac{4}{5}, \frac{2}{3}$

24. Keith's bakery had 252 customers go through its doors last week. This week, that number increased to 378. Express this increase as a percentage.
   a. 26%
   b. 50%
   c. 35%
   d. 12%

25. Sally and seven of her friends get together for their monthly game night to play board games. They choose to start with a game four of them can play, and have to decide who will go first, second, third, and fourth. The other four will either observe the game or play a card game. If anybody can be assigned any position in the order of turns, how many possible orders are there?
   a. 8
   b. 1680
   c. 70
   d. 5040

26. Simplify the following fraction:

$$\frac{\frac{5}{7}}{\frac{9}{11}}$$

a. $\frac{55}{63}$
b. $\frac{7}{1000}$
c. $\frac{13}{15}$
d. $\frac{5}{11}$

27. The following graph compares the various test scores of the top three students in each of these teacher's classes. Based on the graph, which teacher's students had the lowest range of test scores?

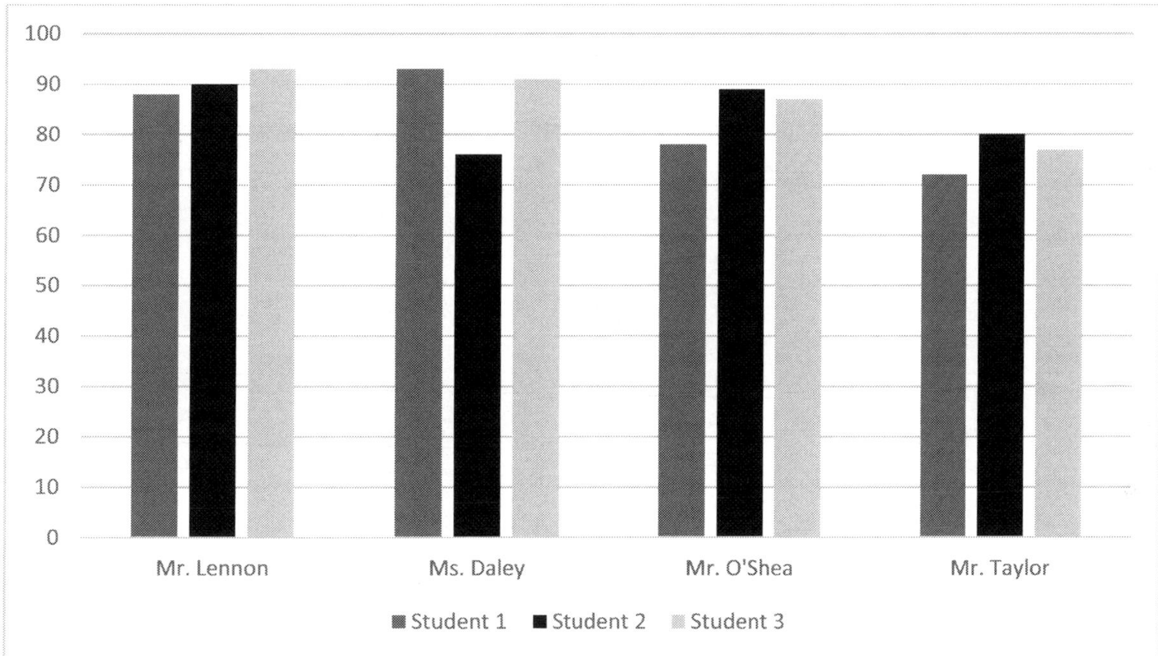

a. Mr. Lennon
b. Mr. O'Shea
c. Mr. Taylor
d. Ms. Daley

28. What is the solution to the following system of equations?

$$x^2 - 2x + y = 8$$
$$x - y = -2$$

a. $(-2, 3)$
b. There is no solution.
c. $(-2, 0)\ (1, 3)$
d. $(-2, 0)\ (3, 5)$

29. Using the following diagram, calculate the total circumference, rounding to the nearest tenth place:

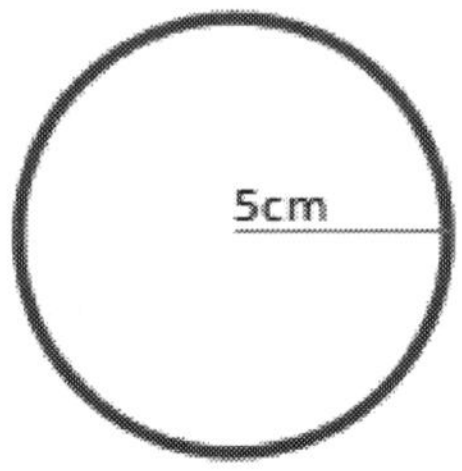

a. 25.0 cm
b. 15.7 cm
c. 78.5 cm
d. 31.4 cm

30. Which measure for the center of a small sample set would be most affected by outliers?
a. Mean
b. Median
c. Mode
d. None of the above

31. A line that travels from the bottom-left of a graph to the upper-right of the graph indicates what kind of relationship between a predictor and a dependent variable?
a. Positive
b. Negative
c. Exponential
d. Logarithmic

32. Which inequality corresponds to the following graph?

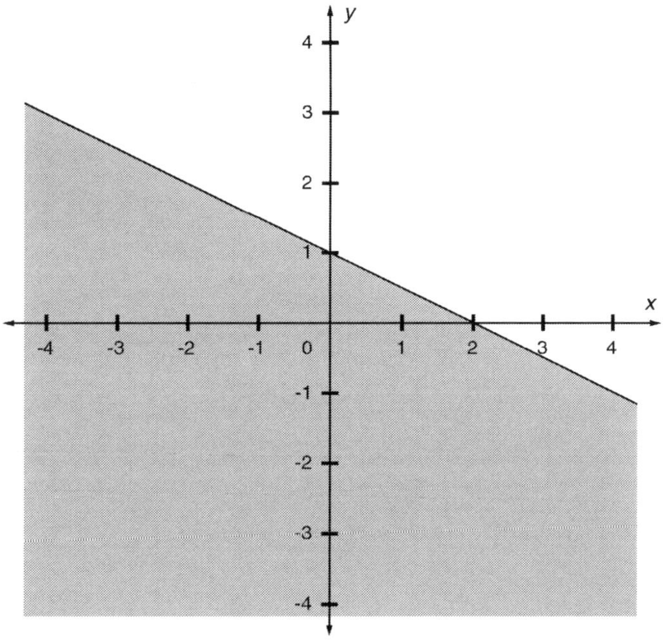

a. $y \leq 1 - \frac{1}{2}x$
b. $y > 1 - \frac{1}{2}x$
c. $y \leq 1 + 2x$
d. $y < 1 + \frac{1}{2}x$

33. Which of the following is the best description of the relationship between Y and X?
a. The data has normal distribution.
b. X and Y have a negative relationship.
c. No relationship
d. X and Y have a positive relationship.

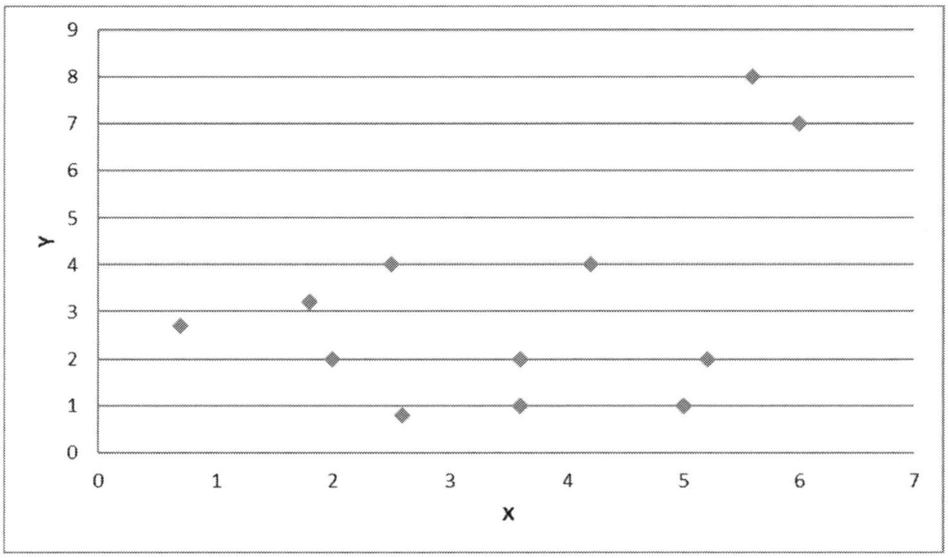

34. Mike set up a simple experiment to measure the rate at which his dough rose over time. He plotted several points onto a blank coordinate plane to track the distance from the top of his pan when he started five hours ago, and the height now. He graphed the highest and lowest two points at (-5, -3) and (0, -1). What is the slope of the line that this makes?

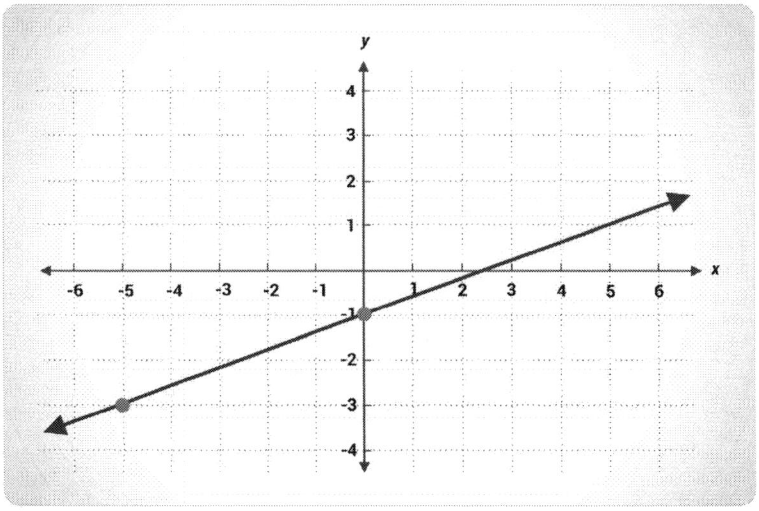

a. 2

b. $\frac{5}{2}$

c. $\frac{1}{2}$

d. $\frac{2}{5}$

35. What is the perimeter of the figure below? Note that the solid outer line is the perimeter.

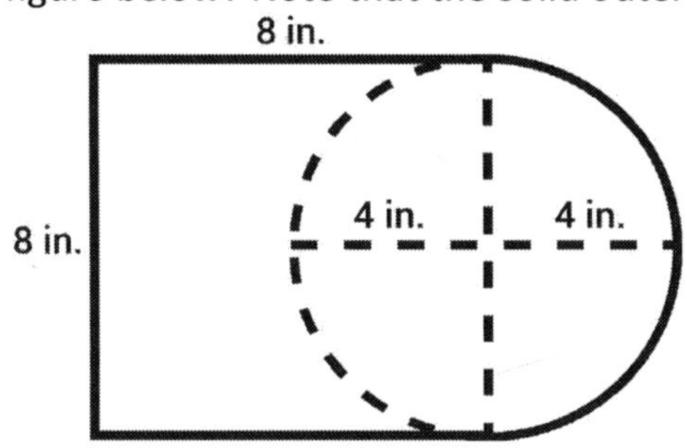

a. 48.565 in
b. 36.565 in
c. 39.78 in
d. 39.565 in

36. Which of the following equations best represents the problem below?

The width of a rectangle is 2 centimeters less than the length. If the perimeter of the rectangle is 44 centimeters, then what are the dimensions of the rectangle?

a. $2l + 2(l - 2) = 44$
b. $(l + 2) + (l + 2) + l = 48$
c. $l \times (l - 2) = 44$
d. $(l + 2) + (l + 2) + l = 44$

37. A ball is drawn at random from a ball pit containing 8 red balls, 7 yellow balls, 6 green balls, and 5 purple balls. What's the probability that the ball drawn is yellow?

a. $\frac{1}{26}$
b. $\frac{19}{26}$
c. $\frac{7}{26}$
d. 1

38. Two cards are drawn from a shuffled deck of 52 cards. What's the probability that both cards are Kings if the first card isn't replaced after it's drawn?

a. $\frac{1}{169}$
b. $\frac{1}{221}$
c. $\frac{1}{13}$
d. $\frac{4}{13}$

39. For a group of 20 men, the median weight is 180 pounds and the range is 30 pounds. If each man gains 10 pounds, which of the following would be true?

a. The median weight will increase, and the range will remain the same.
b. The median weight and range will both remain the same.
c. The median weight will stay the same, and the range will increase.
d. The median weight and range will both increase.

40. For the following similar triangles, what are the values of x and y (rounded to one decimal place)?

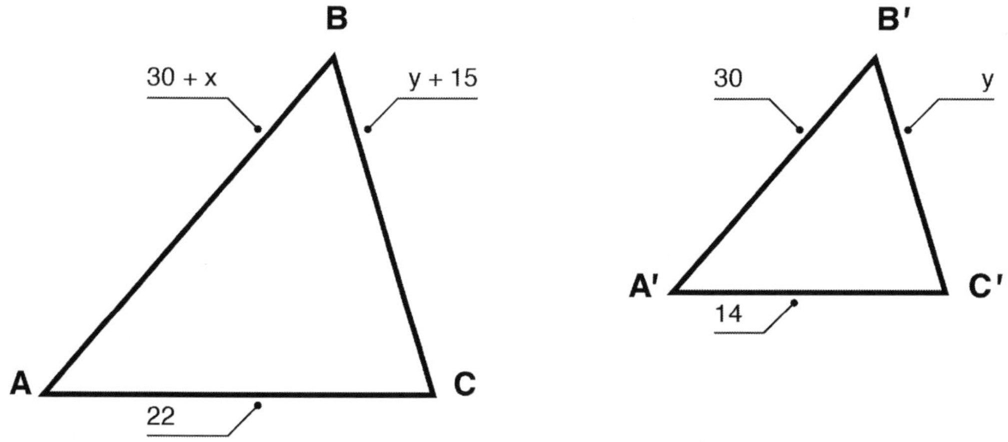

a. $x = 16.5, y = 25.1$
b. $x = 19.5, y = 24.1$
c. $x = 17.1, y = 26.3$
d. $x = 26.3, y = 17.1$

41. A pizzeria owner regularly creates jumbo pizzas, each with a radius of 9 inches. She is mathematically inclined, and wants to know the area of the pizza to purchase the correct boxes and know how much she is feeding her customers. What is the area of the circle, in terms of π, with a radius of 9 inches?
a. $81\ \pi\ \text{in}^2$
b. $18\ \pi\ \text{in}^2$
c. $90\ \pi\ \text{in}^2$
d. $9\ \pi\ \text{in}^2$

42. What is the value of $x^2 - 2xy + 2y^2$ when $x = 2, y = 3$?
a. 8
b. 10
c. 12
d. 14

43. Expand $(2x - 4y)^2$
a. $4x^2 - 16xy + 16y^2$
b. $4x^2 - 8xy + 16y^2$
c. $4x^2 - 16xy - 16y^2$
d. $2x^2 - 8xy + 8y^2$

44. If $x^2 + x - 3 = 0$, then $\left(x + \frac{1}{2}\right)^2 =$

a. $\frac{11}{2}$

b. $\frac{13}{4}$

c. 11

d. $\frac{121}{4}$

45. If $4x - 3 = 5$, then $x =$

a. 1
b. 2
c. 3
d. 4

46. If $x^2 - 2x - 8 = 0$, what value of x makes the equation true?

a. $2 \pm \frac{\sqrt{30}}{2}$

b. $2 \pm 4\sqrt{2}$

c. 1 ± 3

d. $4 \pm \sqrt{2}$

47. Write the expression for six less than three times the sum of twice a number and one.

a. $2x + 1 - 6$
b. $3x + 1 - 6$
c. $3(x + 1) - 6$
d. $3(2x + 1) - 6$

48. Which of the following inequalities is equivalent to $3 - \frac{1}{2}x \geq 2$?

a. $x \geq 2$
b. $x \leq 2$
c. $x \geq 1$
d. $x \leq 1$

49. For which of the following are $x = 4$ and $x = -4$ solutions?

a. $x^2 + 16 = 0$
b. $x^2 + 4x - 4 = 0$
c. $x^2 - 2x - 2 = 0$
d. $x^2 - 16 = 0$

50. Shawna buys $2\frac{1}{2}$ gallons of paint. If she uses $\frac{1}{3}$ of it on the first day, how much does she have left?

a. $1\frac{5}{6}$ gallons

b. $1\frac{1}{2}$ gallons

c. $1\frac{2}{3}$ gallons

d. 2 gallons

Answer Explanations

1. A: Compare each numeral after the decimal point to figure out which overall number is greatest. In Choices *A* (1.43785) and *C* (1.43592), both have the same tenths (4) and hundredths (3). However, the thousandths is greater in Choice *A* (7), so *A* has the greatest value overall.

2. D: By grouping the four numbers in the answer into factors of the two numbers of the question (6 and 12), it can be determined that $(3 \times 2) \times (4 \times 3) = 6 \times 12$. Alternatively, each of the answer choices could be prime factored or multiplied out and compared to the original value. 6×12 has a value of 72 and a prime factorization of $2^3 \times 3^2$. The answer choices respectively have values of 64, 84, 108, 72, and 144 and prime factorizations of 2^6, $2^2 \times 3 \times 7$, $2^2 \times 3^3$, and $2^3 \times 3^2$, so answer *D* is the correct choice.

3. C: The sum total percentage of a pie chart must equal 100%. Since the CD sales take up less than half of the chart and more than a quarter (25%), it can be determined to be 40% overall. This can also be measured with a protractor. The angle of a circle is 360°. Since 25% of 360° would be 90° and 50% would be 180°, the angle percentage of CD sales falls in between; therefore, it would be Choice *C*.

4. B: Since $850 is the price *after* a 20% discount, $850 represents 80% of the original price. To determine the original price, set up a proportion with the ratio of the sale price (850) to original price (unknown) equal to the ratio of sale percentage (where x represents the unknown original price):

$$\frac{850}{x} = \frac{80}{100}$$

To solve a proportion, cross multiply the numerators and denominators and set the products equal to each other:

$$(850)(100) = (80)(x)$$

Multiplying each side results in the equation:

$$85{,}000 = 80x$$

To solve for x, divide both sides by 80:

$$\frac{85{,}000}{80} = \frac{80x}{80}$$

$$x = 1062.5$$

Remember that x represents the original price. Subtracting the sale price from the original price ($\$1062.50 - \$850$) indicates that Frank saved $212.50.

5. C: Division in scientific notation can be solved by grouping the first terms together and grouping the tens together. The first terms can be divided, and the tens terms can be simplified using the rules for exponents. The initial expression becomes 0.4×10^4. This is not in scientific notation because the first number is not between 1 and 10. Shifting the decimal and subtracting one from the exponent, the answer becomes 4.0×10^3.

6. C: The value of the factorial $n!$ is found by the product all integers from 1 to n. The factorial $5!$ is evaluated as:

$$5! = 5 \times 4 \times 3 \times 2 \times 1 = 120$$

Thus, the expression is equal to 120.

7. A: The Pythagorean theorem states that for right triangles $c^2 = a^2 + b^2$, with c being the side opposite the 90° angle. Substituting 24 as a and 36 as b, the equation becomes:

$$c^2 = 24^2 + 36^2$$

$$576 + 1296 = 1872$$

The last step is to square both sides to remove the exponent:

$$c = \sqrt{1872} = 43.3$$

8. C: The first step in solving this problem is expressing the result in fraction form. Separate this problem first by solving the division operation of the last two fractions. When dividing one fraction by another, invert or flip the second fraction and then multiply the numerator and denominator.

$$\frac{7}{10} \times \frac{2}{1} = \frac{14}{10}$$

Next, multiply the first fraction with this value:

$$\frac{3}{5} \times \frac{14}{10} = \frac{42}{50}$$

Decimals are expressions of 1, or 100%, so multiply both the numerator and denominator by 2 to get the fraction as an expression of 100.

$$\frac{42}{50} \times \frac{2}{2} = \frac{84}{100}$$

In decimal form, this would be expressed as 0.84.

9. A: ○. The core of the pattern consists of 4 items: ▲○○□. Therefore, the core repeats in multiples of 4, with the pattern starting over on the next step. The closest multiple of 4 to 42 is 40. Step 40 is the end of the core (□), so step 41 will start the core over (▲), and step 42 is ○.

10. D: This problem can be solved by setting up a proportion involving the given information and the unknown value. The proportion is:

$$\frac{21\ pages}{4\ nights} = \frac{140\ pages}{x\ nights}$$

Solving the proportion by cross-multiplying, the equation becomes $21x = 4 \times 140$, where $x = 26.67$. Since it is not an exact number of nights, the answer is rounded up to 27 nights. Twenty-six nights would not give Sarah enough time.

11. B: Using the conversion rate, multiply the projected weight loss of 25 lbs. by 0.45 $\frac{kg}{lb}$ to get the amount in kilograms (11.25 kg).

12. D: First, subtract $1437 from $2334.50 to find Johnny's monthly savings; this equals $897.50. Then, multiply this amount by 3 to find out how much he will have (in three months) before he pays for his vacation: this equals $2692.50. Finally, subtract the cost of the vacation ($1750) from this amount to find how much Johnny will have left: $942.50.

13. B: Dividing by 98 can be approximated by dividing by 100, which would mean shifting the decimal point of the numerator to the left by 2. The result is 4.2, which rounds to 4. Sarah may have about 4 full sets of figurines at the most.

14. B: The value of x within the radical can be solved by first isolating the radical expression:

$$\sqrt{3x+5} - 2 = 1$$

$$\sqrt{3x+5} = 3$$

Next, the radical can be cancelled out by squaring both sides of the equation:

$$\left(\sqrt{3x+5}\right)^2 = 3^2$$

$$3x + 5 = 9$$

Finally, the equation can be solved in terms of x by subtracting 5 from both sides and dividing by 3:

$$3x + 5 - 5 = 9 - 5$$

$$3x = 4$$

$$\frac{3x}{3} = \frac{4}{3}$$

$$x = \frac{4}{3}$$

Thus, the value of x that makes the equality true is $\frac{4}{3}$.

15. C: The formula for the perimeter of a rectangle is $P = 2L + 2W$, where P is the perimeter, L is the length, and W is the width. The first step to find width is to substitute all of the data into the formula:

$$36 = 2(12) + 2W$$

Simplify by multiplying $2x12$:

$$36 = 24 + 2W$$

Simplifying this further by subtracting 24 on each side, which gives:

$$36 - 24 = 24 - 24 + 2W$$

$$12 = 2W$$

Divide by 2:

$$6 = W$$

The width of the decoration's base is 6 cm. Remember to test this answer by substituting this value into the original formula:

$$36 = 2(12) + 2(6)$$

16. D: To find the average of a set of values, add the values together and then divide by the total number of values. In this case, include the unknown value of what Dwayne needs to score on his next test, in order to solve it.

$$\frac{78 + 92 + 83 + 97 + x}{5} = 90$$

Add the unknown value to the new average total, which is 5. Then multiply each side by 5 to simplify the equation, resulting in:

$$78 + 92 + 83 + 97 + x = 450$$

$$350 + x = 450$$

$$x = 100$$

Dwayne would need to get a perfect score of 100 in order to get an average of at least 90.

Test this answer by substituting back into the original formula.

$$\frac{78 + 92 + 83 + 97 + 100}{5} = 90$$

17. D: For an even number of total values, the *median* is calculated by finding the *mean* or average of the two middle values once all values have been arranged in ascending order from least to greatest. In this case, $(92 + 83) \div 2$ would equal the median 87.5, Choice *D*.

18. C: Three squared subtracted from the product of the square roots of 36 and 16 is the same as:

$$\left(\sqrt{36} \times \sqrt{16}\right) - 3^2$$

Follow the *order of operations* in order to solve this problem. Solve inside the parentheses first, followed by multiplication and then subtraction:

$$(6 \times 4) - 9$$

This is equal to $24 - 9$, or 15, Choice *C*.

19. D: $\frac{1}{12}$. The probability of picking the winner of the race is $\frac{1}{4}$, or $\left(\frac{number\ of\ favorable\ outcomes}{number\ of\ total\ outcomes}\right)$. Assuming the winner was picked on the first selection, three horses remain from which to choose the runner-up (these are dependent events). Therefore, the probability of picking the runner-up is $\frac{1}{3}$. To determine the probability of multiple events, the probability of each event is multiplied:

$$\frac{1}{4} \times \frac{1}{3} = \frac{1}{12}$$

20. B: The square of the difference between 25 and 21 divided by 2 and added to 4 times 7 is equal to:

$$4 \times 7 + (25 - 21)^2 \div 2$$

To solve this correctly, keep in mind the order of operations with the mnemonic PEMDAS (Please Excuse My Dear Aunt Sally). This stands for Parentheses, Exponents, Multiplication, Division, Addition, Subtraction. Taking it step by step, solve inside the parentheses first:

$$4 \times 7 + (4)^2 \div 2$$

Then, apply the exponent:

$$4 \times 7 + 16 \div 2$$

Multiplication and division are both performed next:

$$28 + 8 = 36$$

Addition and subtraction are done last. The solution is 36.

21. C: Kimberley worked 4.5 hours at the rate of \$10/h and 1 hour at the rate of \$12/h. The problem states that her pay is rounded to the nearest hour, so the 4.5 hours would round up to 5 hours at the rate of \$10/h.

$$(5\ h)(\$10/h) + (1\ h)(\$12/h) = \$50 + \$12 = \$62.$$

22. D:

$9x + x - 7 = 16 + 2x$	Combine $9x$ and x.
$10x - 7 = 16 + 2x$	
$10x - 7 + 7 = 16 + 2x + 7$	Add 7 to both sides to remove (-7).
$10x = 23 + 2x$	
$10x - 2x = 23 + 2x - 2x$	Subtract 2x from both sides to move it to the other side of the equation.
$8x = 23$	
$\frac{8x}{8} = \frac{23}{8}$	Divide by 8 to get x by itself.
$x = \frac{23}{8}$	

23. C: The first step is to depict each number using decimals:

$$\frac{91}{100} = 0.91$$

Dividing the numerator by the denominator of $\frac{4}{5}$ to convert it to a decimal yields 0.80, while $\frac{2}{3}$ becomes 0.66 recurring. Rearrange each expression in ascending order, as found in Choice *C*.

24. B: First, calculate the difference between the larger value and the smaller value.

$$378 - 252 = 126$$

To calculate this difference as a percentage of the original value, and thus calculate the percentage *increase*, divide 126 by 252, then multiply by 100 to reach the percentage 50%, Choice *B*.

25. B: The number of arrangements, or permutations, of players where order matters is calculated by the following equation, where $P(n, r)$ is the number of permutations, n is the total number of people, and k is the number of people selected:

$$P(n, r) = \frac{n!}{(n - r)!}$$

Since there are eight people and four can be selected, $n = 8$ and $r = 4$:

$$P(8, 4) = \frac{8!}{(8 - 4)!} = \frac{8!}{4!}$$

Dividing factorials is simple because the lesser factorial cancels out part of the greater factorial:

$$P(8, 4) = \frac{8 \times 7 \times 6 \times 5 \times 4 \times 3 \times 2 \times 1}{4 \times 3 \times 2 \times 1} = 8 \times 7 \times 6 \times 5 = 1680$$

Thus, the number of arrangements of 4 players out of 8 people is 1680.

26. A: First simplify the larger fraction by separating it into two. When dividing one fraction by another, remember to *invert* the second fraction and multiply the two as follows:

$$\frac{5}{7} \times \frac{11}{9}$$

The resulting fraction of $\frac{55}{63}$ cannot be simplified further, so this is the answer to the problem.

27. A: To calculate the range in a set of data, subtract the highest value with the lowest value. In this graph, the range of Mr. Lennon's students is 5, which can be seen physically in the graph as having the smallest difference compared with the other teachers between the highest value and the lowest value.

28. D: This system of equations involves one quadratic function and one linear function, as seen from the degree of each equation. One way to solve this is through substitution. Solving for y in the second equation yields $y = x + 2$. Plugging this equation in for the y of the quadratic equation yields:

$$x^2 - 2x + x + 2 = 8$$

Simplifying the equation, it becomes $x^2 - x + 2 = 8$. Setting this equal to zero and factoring, it becomes:

$$x^2 - x - 6 = 0 = (x - 3)(x + 2)$$

Solving these two factors for x gives the zeros $x = 3, -2$. To find the y-value for the point, each number can be plugged in to either original equation. Solving each one for y yields the points $(3, 5)$ and $(-2, 0)$.

29. D: To calculate the circumference of a circle, use the formula $2\pi r$, where r equals the radius or half of the diameter of the circle and $\pi = 3.14$. Substitute the given information, $2\pi 5 = 31.4$, Choice *D*.

30. A: Mean. An outlier is a data value that's either far above or below the majority of values in a sample set. The mean is the average of all values in the set. In a small sample, a very high or low number could greatly change the average. The median is the middle value when arranged from lowest to highest. Outliers would have no more of an effect on the median than any other value. Mode is the value that repeats most often in a set. Assuming that the same outlier doesn't repeat, outliers would have no effect on the mode of a sample set.

31. A: This vector indicates a positive relationship. A negative relationship would show points traveling from the top-left of the graph to the bottom-right. Exponential and logarithmic functions aren't linear (they don't create a straight line), so these options could have been immediately eliminated.

32. A: The line graphed by the inequality has an x-intercept of 2 and a y-intercept of 1, which gives the equation a slope of $-\frac{1}{2}x$, found by taking the change in y, -1, and dividing it by the change in x, 2. So, the equation of the line is $y = 1 - \frac{1}{2}x$ Because the shaded region lies below the equation, and the line is solid, the inequality must contain all the points on and below the line: $y \leq 1 - \frac{1}{2}x$.

33. C: There is no verifiable relationship between the two variables. While it may seem to have somewhat of a positive correlation because of the last two data points: (5.6, 8) and (6, 7), you must also take into account the two data points before those (5, 1) and (5.2, 2) that have low Y values despite high X values. Data with a normal distribution (Choice *A*) has an arc to it. This data does not.

34. D: The slope is given by the change in y divided by the change in x. Specifically, it's:

$$slope = \frac{y_2 - y_1}{x_2 - x_1}$$

The first point is (-5, -3), and the second point is (0, -1). Work from left to right when identifying coordinates. Thus the point on the left is point 1 (-5, -3) and the point on the right is point 2 (0, -1).

Now we need to just plug those numbers into the equation:

$$slope = \frac{-1 - (-3)}{0 - (-5)}$$

It can be simplified to:

$$slope = \frac{-1 + 3}{0 + 5}$$

$$slope = \frac{2}{5}$$

35. B: The figure is composed of three sides of a square and a semicircle. The sides of the square are simply added: $8 + 8 + 8 = 24$ inches. The circumference of a circle is found by the equation $C = 2\pi r$. The radius is 4 inches, so the circumference of the circle is 25.132 inches. Only half of the circle makes up the outer border of the figure (part of the perimeter) so half of 25.132 inches is 12.566 inches. Therefore, the total perimeter is: $24 \text{ in} + 12.566 \text{ in} = 36.566 \text{ in}$. The other answer choices use the incorrect formula or fail to include all of the necessary sides.

36. A: The first step is to determine the unknown, which is in terms of the length, l.

The second step is to translate the problem into the equation using the perimeter of a rectangle, $P = 2l + 2w$. The width is the length minus 2 centimeters. The resulting equation is $2l + 2(l - 2) = 44$. The equation can be solved as follows:

$2l + 2l - 4 = 44$	Apply the distributive property on the left side of the equation
$4l - 4 = 44$	Combine like terms on the left side of the equation
$4l = 48$	Add 4 to both sides of the equation
$l = 12$	Divide both sides of the equation by 4

The length of the rectangle is 12 centimeters. The width is the length minus 2 centimeters, which is 10 centimeters. Checking the answers for length and width forms the following equation:

$$44 = 2(12) + 2(10)$$

The equation can be solved using the order of operations to form a true statement: $44 = 44$. Because the equation correctly models the formula for perimeter and has a valid solution, Choice *A* is correct.

37. C: The sample space is made up of $8 + 7 + 6 + 5 = 26$ balls. The probability of pulling each individual ball is $\frac{1}{26}$. Since there are 7 yellow balls, the probability of pulling a yellow ball is $\frac{7}{26}$.

38. B: For the first card drawn, the probability of a King being pulled is $\frac{4}{52}$. Since this card isn't replaced, if a King is drawn first the probability of a King being drawn second is $\frac{3}{51}$. The probability of a King being drawn in both the first and second draw is the product of the two probabilities: $\frac{4}{52} \times \frac{3}{51} = \frac{12}{2652}$. This fraction, when divided by 12, equals $\frac{1}{221}$.

39. A: If each man gains 10 pounds, every original data point will increase by 10 pounds. Therefore, the man with the original median will still have the median value, but that value will increase by 10. The smallest value and largest value will also increase by 10 and, therefore, the difference between the two won't change. The range does not change in value and, thus, remains the same.

40. C: Because the triangles are similar, the lengths of the corresponding sides are proportional. Therefore:

$$\frac{30+x}{30} = \frac{22}{14} = \frac{y+15}{y}$$

This results in the equation $14(30 + x) = 22 \times 30$ which, when solved, gives $x = 17.1$. The proportion also results in the equation $14(y + 15) = 22y$ which, when solved, gives $y = 26.3$.

41. A: The formula for the area of the circle is πr^2 and 9 squared is 81. Choice *B* is not the correct answer because that is 2×9. Choice *C* is not the correct answer because that is 9×10. Choice *D* is not the correct answer because that is simply the value of the radius.

42. B: Start with the original equation: $x^2 - 2xy + 2y^2$, then replace each instance of x with a 2, and each instance of y with a 3 to get:

$$2^2 - 2 \times 2 \times 3 + 2 \times 3^2 = 4 - 12 + 18 = 10$$

43. A: To expand a squared binomial, it's necessary to use the FOIL (*First, Inner, Outer, Last)* Method. Start with the original equation:

$(2x \times 2x)$	+	$(2x \times -4y)$	+	$(2x \times -4y)$	+	$(-4y \times -4y)$
First		Outer		Inner		Last

$$= 4x^2 - 8xy - 8xy + 16y^2$$

$$= 4x^2 - 16xy + 16y^2$$

44. B: The first step is to use the quadratic formula on the first equation ($x^2 + x - 3 = 0$) to solve for x. In this case, a is 1, b is 1, and c is -3, yielding:

$$x = \frac{-b \pm \sqrt{b^2 - 4ac}}{2a}$$

$$x = \frac{-1 \pm \sqrt{1 - 4 \times 1(-3)}}{2}$$

$$x = \frac{-1}{2} \pm \frac{\sqrt{13}}{2}$$

Therefore, $x + \frac{1}{2}$, which is in our second equation, equals $\pm\frac{\sqrt{13}}{2}$. We are looking for $\left(x + \frac{1}{2}\right)^2$ though, so we square the $\pm\frac{\sqrt{13}}{2}$. Doing so causes the $\pm$ to cancel and it leaves $\left(\frac{\sqrt{13}}{2}\right)^2 = \frac{13}{4}$.

45. B: Add 3 to both sides to get $4x = 8$. Then divide both sides by 4 to get $x = 2$.

46. C: The equation is a polynomial, so the first step is to try factoring the equation. The numbers needed are those that add to -2 and multiply to -8. The difference between 2 and 4 is 2. Their product is 8, and -4 and 2 will work. Therefore:

$$x^2 - 2x - 8 = (x - 4)(x + 2)$$

The latter has roots 4 and -2, which also translates as 1 ± 3.

47. D: The expression is three times the sum of twice a number and 1, which is $3(2x + 1)$. Then, 6 is subtracted from this expression.

48. B: To simplify this inequality, subtract 3 from both sides to get $-\frac{1}{2}x \geq -1$. Then, multiply both sides by -2 (remembering this flips the direction of the inequality) to get $x \leq 2$.

49. D: There are two ways to approach this problem. Each value can be substituted into each equation. A can be eliminated, since $4^2 + 16 = 32$. Choice *B* can be eliminated, since:

$$4^2 + 4 \times 4 - 4 = 28$$

Choice *C* can be eliminated, since:

$$4^2 - 2 \times 4 - 2 = 6$$

But, plugging in either value into $x^2 - 16$, which gives:

$$(\pm 4)^2 - 16 = 16 - 16 = 0$$

50. C: If she has used 1/3 of the paint, she has 2/3 remaining. $2\frac{1}{2}$ gallons are the same as $\frac{5}{2}$ gallons. The calculation is:

$$\frac{2}{3} \times \frac{5}{2} = \frac{5}{3} = 1\frac{2}{3} \text{ gallons}$$

Dear CHSPE Test Taker,

We would like to start by thanking you for purchasing this study guide for your CHSPE exam. We hope that we exceeded your expectations.

Our goal in creating this study guide was to cover all of the topics that you will see on the test. We also strove to make our practice questions as similar as possible to what you will encounter on test day. With that being said, if you found something that you feel was not up to your standards, please send us an email and let us know.

We would also like to let you know about other books in our catalog that may interest you.

SAT

This can be found on Amazon: amazon.com/dp/1628457376

ACT

amazon.com/dp/1628459468

ACCUPLACER

amazon.com/dp/1628459344

College Placement

amazon.com/dp/1628459506

We have study guides in a wide variety of fields. If the one you are looking for isn't listed above, then try searching for it on Amazon or send us an email.

Thanks Again and Happy Testing!
Product Development Team
info@studyguideteam.com

FREE Test Taking Tips DVD Offer

To help us better serve you, we have developed a Test Taking Tips DVD that we would like to give you for FREE. **This DVD covers world-class test taking tips that you can use to be even more successful when you are taking your test.**

All that we ask is that you email us your feedback about your study guide. Please let us know what you thought about it – whether that is good, bad or indifferent.

To get your **FREE Test Taking Tips DVD**, email freedvd@studyguideteam.com with "FREE DVD" in the subject line and the following information in the body of the email:

a. The title of your study guide.

b. Your product rating on a scale of 1-5, with 5 being the highest rating.

c. Your feedback about the study guide. What did you think of it?

d. Your full name and shipping address to send your free DVD.

If you have any questions or concerns, please don't hesitate to contact us at freedvd@studyguideteam.com.

Thanks again!

Made in the USA
Las Vegas, NV
21 January 2021

16275191R00092